Automotive Safety Handbook

Other SAE titles of interest:

Role of the Seat in Rear Crash Safety
by David C. Viano
(Order No. R-317)

Fundamentals of Crash Sensing in Automotive Air Bag Systems
by Ching-Yao Chan
(Order No. R-217)

SAE Active Safety Technology Collection on CD-ROM
(Order No. ACTSAFE2002)

For more information or to order a book, contact SAE at
400 Commonwealth Drive, Warrendale, PA 15096-0001;
phone (724) 776-4970; fax (724) 776-0790;
e-mail CustomerService@sae.org;
website http://store.sae.org.

Automotive Safety Handbook

Ulrich Seiffert

and

Lothar Wech

**Professional
Engineering
Publishing**

SAEInternational™

Warrendale, Pa.

London, UK

For permission and licensing requests, contact:

SAE Permissions
400 Commonwealth Drive
Warrendale, PA 15096-0001 USA
E-mail: permissions@sae.org
Tel: 724-772-4028
Fax: 724-772-4891

Library of Congress Cataloging-in-Publication Data
Seiffert, Ulrich.
  Automotive safety handbook / Ulrich Seiffert and Lothar Wech.
  p. cm.
  Includes bibliographical references.
  ISBN 0-7680-0912-X
  1. Automobiles—Safety appliances—Handbooks, manuals, etc.
  2. Automobiles—Testing—Handbooks, manuals, etc. I. Wech, Lothar. II. Title.

  TL159.5 .S45 2003
  629.2'31—dc21   2003052646

This edition published by SAE International and co-published in the UK by Professional Engineering Publishing Limited.

SAE
400 Commonwealth Drive
Warrendale, PA 15096-0001 USA
E-mail: CustomerService@sae.org
Tel: 877-606-7323 (USA/Canada)
        724-776-4970 (outside USA)
Fax: 724-776-1615

Professional Engineering Publishing Limited
Northgate Avenue, Bury St Edmunds
Suffolk, IP32 6BW, UK
E-mail: orders@pepublishing.com
Tel: +44 (0) 1284 724 384
Fax: +44 (0) 1284 718 692
http://www.pepublishing.com

Copyright © 2003 SAE International
SAE ISBN 0-7680-0912-X;
Professional Engineering Publishing ISBN 1-86058-346-6
SAE Order No. R-325
Printed in the United States of America.

Preface

This book on automotive safety describes the relevant development of safety for passenger cars. Because of our long-standing experience in the field of automotive safety (i.e., both authors have worked for more than two decades in this field), basic relationships and new developments related to accident avoidance and mitigation of injuries are described. Included in discussions are driver support systems, chassis, lights, body and interior design, restraint systems, biomechanics, dummies, accident simulation tests, pedestrian protection, and compatibility.

This book is suitable for those who are interested in safety engineering, and for students and experts who are interested not only in details but also in the broad perspective of vehicle safety.

Special thanks are extended to the many persons and companies supporting this book with relevant material, especially Autoliv, Bosch, Continental, DaimlerChrysler, and Volkswagen.

Table of Contents

1.
Introduction

Today, "safety and security" in all our activities has become more and more a basic element in our day-to-day experiences. There are several reasons for this development. The largest influence is the increasing number of people in the world. As a consequence, more resources are being used for daily living than the Earth might be able to provide. Other parameters are the rapid rate of change in the world (e.g., modern communication techniques), the use of all kinds of energy, increasing social differences, the consumption of fossil fuels, local and global accidents, and crime. This list could easily be made much longer. Because information systems are available 24 hours per day worldwide, knowledge about these factors and dramatic events is much greater today than in the past.

For this reason, it is not surprising that in marketing questionnaires related to customer wishes for new cars, when customers are asked which performance characteristic is most important in their decision-making processes for buying a new car, safety is on the top of the list. Applied to cars, this means that customers desire not only safe and comfortable driving without breakdowns, but also integrated information and communication systems and to be as safe as possible in the event of an accident. Figure 1.1 shows that vehicle safety is one of the important areas in customers' overall requirements for cars [1-1].

This book describes the most important areas of vehicle safety in passenger cars. To work in the field of automotive safety is one of the most positive activities to which an engineer can commit during his or her career.

1.1 Reference

1-1. Seiffert, U. Fahrzeugsicherheit VDI Verlag, Düsseldorf, 1992, ISBN 3-18-401264-6.

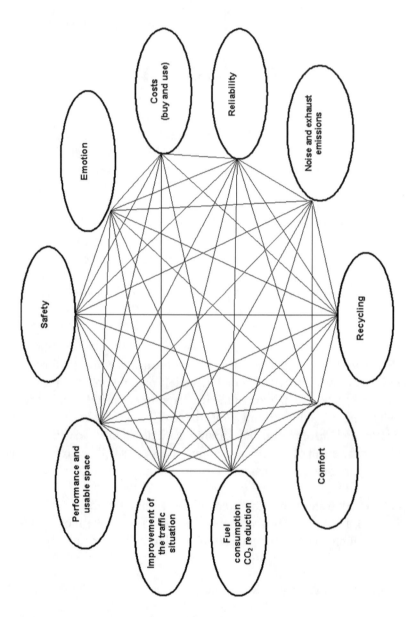

Figure 1.1 Conflicting demands by vehicle buyers.

2.
Definitions

There are many definitions for vehicle safety. One of the best is based on the statement made in 1955 by the German university professor, Professor Koeßler: "The motor-vehicle has the duty to transport humans and goods from place A to place B, as safely, as quickly, and as comfortably as possible." In modern life, we would add to this definition "...and as efficiently and environmentally friendly as possible." Figure 2.1 illustrates a more scientific perspective of vehicle safety.

Vehicle safety can be subdivided into two main areas:

1. Accident avoidance
2. Mitigation of injuries

Some additional definitions related to vehicle safety are as follows:

- **Accident avoidance.** (Colloquial = Active safety.) All measures that serve to prevent accidents.

- **Mitigation of injuries.** (Colloquial = Passive safety.) All measures that help to reduce injuries during accidents.

- **External safety**. Design of the external parts of the vehicle to reduce injuries in the event of a collision with an external collision partner.

- **Interior safety.** Design of the vehicle interior parts to prevent additional injuries in the event of contact with an occupant of the vehicle.

- **Restraint systems**. Vehicle components (e.g., seat belts, airbags) that specifically influence the relative movement of occupants in relation to the vehicle.

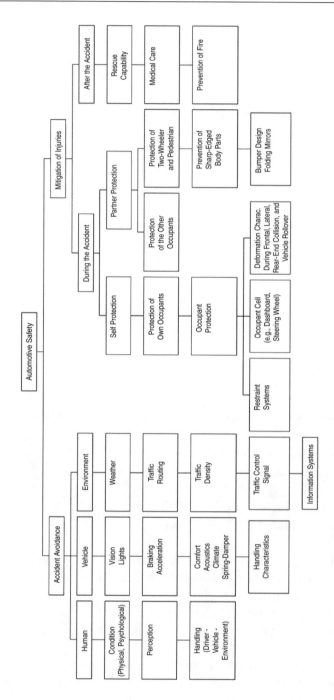

Figure 2.1 Automotive safety.

- **Smart restraints, sensors, and actuators**. These apply to occupant detection and pre-crash evaluation.

- **Primary collision.** Collision of the vehicle with another obstacle.

- **Secondary collision**. Collision of the occupant with vehicle parts.

- **Active devices**. Safety features that must be activated manually for use (e.g., active seat belts).

- **Passive devices.** Safety and restraint systems that in an accident are activated automatically (e.g., airbags, pretensioners in seat belts, automatic height-adjusted headrests).

The two primary safety fields, accident avoidance and mitigation of injuries, often are called active safety and passive safety, respectively, in the public domain. Although these phrases described the situation clearly until the end of the 1970s, they are no longer valid today. Many passive safety systems, such as airbags and seat-belt pretensioners, can react in a very active manner, and the two areas are moving closer together. For example, near-distance radar sensors could be used for several purposes, including the inflation of airbags and the activation of seat-belt pretensioners.

In the field of accident avoidance, three main influences are found:

1. The human being
2. The technical features of a car
3. The environment

In many studies, human beings are cited as the main cause of accidents. If this statement is examined in more detail, you will find strong influences from other areas (e.g., comfort, noise, information technology [cellular phones], and the man-machine interface) that might change this high number. In any case, the driver is influenced by his or her mental and physical conditions, as well as any drugs and/or alcohol that he or she may have consumed. With respect to the vehicle, the general handling characteristics, driver support systems (e.g., antilock brake system [ABS], automatic cruise control [ACC], electronic stability program [ESP], power steering, automatic transmission, and information systems), field of view, lighting, and comfort level (e.g., noise and vibration conditions, heating and air conditioning levels) influence the behavior of the vehicle [2-1]. To some extent, the road (i.e., layout, road surface, and traffic signals) and the weather also play important roles. Some functions in the field of accident avoidance are being combined with items from the field of mitigation of injuries. One example is sensor fusion, where sensors are used for both accident avoidance and as the pre-crash signal for airbags (i.e., obstacle detection). Information systems can help to reduce the occurrence of accidents and can notify an ambulance and the police automatically if a severe accident occurs.

The field of mitigation of injuries can be subdivided into two main groups:

1. Protection during the crash
2. Protection after the accident

These two groups can be further subdivided into the protection of the vehicle occupants in both vehicles in a multi-vehicle collision and, in a single-vehicle crash, the protection of the occupants of that vehicle. Another important field is post-crash safety, with several requirements such as the capability to open the doors of the vehicle without tools and the prevention of fire.

2.1 Reference

2-1. Braess, H.-H. and Seiffert, U. *Handbuch Kraftfahrzeugtechnik*, Vieweg-Verlag, Wiesbaden, Germany, 2001.

3.

Driving Forces for Increased Vehicle Safety

Many factors and activities have helped to reduce the number of accidents on all types of roads. The following are examples, not listed by priority:

- Customer demand
- Science
- Public demand
- Technology
- Government legislation
- Product liability
- Consumer information
- Competition among car manufacturers
- Automotive press reports

Legislation certainly has a strong influence on safety. In Germany, with the development of the first cars, a proposal for a law about liability for the use of vehicles was introduced in 1909. In 1832, England introduced requirements for a steam-powered bus with respect to performance during accidents. A strong positive effect of legislation surfaced in rule-making activities in the United States during the middle of the 1960s and during the International Experimental Safety Vehicle Conference, where safety became a competitive item for medical and technical scientists, legislators, and automotive engineers.

A second effect was the activity initiated by consumer advocates such as Ralph Nader in the United States, and the increasing number of consumer information reports with respect to the performance of vehicles in the field and in accident simulation tests, such as the crashworthiness rating report issued by the Insurance Institute for Highway Safety (IIHS) [3-1], the Highway Loss Data Institute [3-2], and the worldwide New Car Assessment Program (NCAP) crash tests [3-3]. Table 3.1 shows an overview of the various NCAP activities in Europe, the United States, Japan, and Australia, together with those of IIHS.

TABLE 3.1
WORLDWIDE NCAP TEST [3-5]

	Euro NCAP	U.S. NCAP	IIHS	A NCAP	J NCAP
Rigid wall full frontal impact	—	56 km/h H III H III	—	56 km/h H III H III	55 km/h H III H III
Offset deformable barrier (EEVC) frontal impact	64 km/h H III H III P3 P 1 1/2	—	64 km/h H III	64 km/h H III H III P3 P 1 1/2	64 km/h H III H III
Mobile barrier side impact	50 km/h EEVC barrier EuroSID I P11/2 P3	62 km/h crab barrier SID SID	—	50 km/h EEVC barrier EuroSID I P11/2 P3	50 km/h EEVC barrier EuroSID I
Side pole impact	29 km/h flying floor EuroSID I	—	—	—	—
Pedestrian bodyform impacts	40 km/h adult head child head upper leg lower leg	—	—	—	—

The information available to the public about the performance in the preceding tests and about other criteria is provided not only by government organizations [3-4] but by automotive magazines [3-5]—even by the German auto enthusiasts' magazine, *Auto, Motor, and Sport* in Germany. Interested consumers also can find the results on various websites. Because of consumer interest for information in addition to the data that are already available, more information related to vehicle performance with respect to accident avoidance is under discussion for distribution. In addition to the two parameters already mentioned, a third may be the most important one. This is the increasing interest by customers, who rank safety features when buying new cars as "extremely" or "very" important. When customers were asked about the importance of car safety, the percentage of people who showed high interest in vehicle safety increased from 64% in 1981 to 84% in 1999 in the United States [3-6]. In Germany, companies such as Audi and Volkswagen have published figures that state that 85% of people who were asked mentioned safety as the top issue. In the early 1990s, safety became a competitive item among automobile manufacturers. This is one of the best things that could happen for both the consumer and the field of safety. The increasing interest of all groups mentioned may be connected with the longing for a secure and safe life.

Although not discussed in public intensively, product liability legislation also has had some important effects related to the design of cars. In product liability cases, the state-of-the-art defense leads to the fact that safety features, which are introduced by one manufacturer, create pressure on other manufacturers to install these new safety devices as standard equipment on all cars. The high level of attention to road safety also has increased the number and the quality of accident analyses performed by different institutions in various countries. This is true not only for the government organizations but also for insurance companies, scientific and medical organizations, and some of the major car companies.

3.1 References

3-1. Crashworthiness rating reports, various issues, sporadically published by Insurance Institute for Highway Safety, Arlington, VA, USA.

3-2. Highway Loss Data Institute, www.carsafety.org.

3-3. Klanner, W. "Status Report and Future Development of the Euro NCAP Program," Proceedings 17 Experimental Safety Vehicles (ESV) Conference, 2001.

3-4. Crashworthiness rating reports, various issues, sporadically published by National Highway Traffic Safety Administration (NHTSA) and the Federal Motor Vehicle Safety Standards (FMVSS), Washington, DC, United States.

3-5.   "Safety Steps in the Spotlight," *Automotive News*, March 6, 2000.

3-6.   Insurance Institute for Highway Safety, Special Issue: "Vehicle Compatibility in Crashes," October 1999.

4.
Safety Legislation

The different government legislative institutions that are responsible for the improvement of traffic safety are important because they combine three elements:

1. Traffic routing (e.g., roads and traffic signals)
2. The education of traffic participants
3. Vehicle performance

The most important requirements do not describe the design but the performance criteria that must be fulfilled in defined tests. Only such rules support the creativity of engineers and the competition of different ideas.

With respect to automotive legislation on a worldwide basis, there are different starting points for various countries. For example, the meaning set forth by the responsible rule-making bodies in the United States was that traffic participants, especially the driver, could be educated to only a limited extent; thus, vehicle occupants must be protected in the event of an accident caused by the vehicle itself (design).

The second priority was given to measures for accident avoidance. Contrary to the U.S. perspective, European legislators gave much more responsibility to the driver because the focal point was the prevention of accidents. In the beginning and until the mid-1960s, the number of accidents with severe and fatal injuries reached such a high level that the American government defined in November 1966, with the Motor Vehicle Safety Act [4-1], intensive requirements for automobiles. Europe and other countries followed. Meanwhile, several of the 100 requirements were introduced worldwide (Figure 4.1).

Number

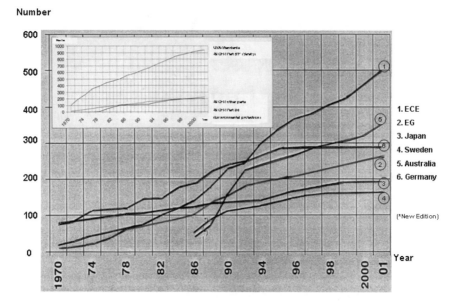

Figure 4.1 Number of rules for vehicles in various countries.
(Source: Volkswagen AG.)

In Europe, the situation with respect to the rule-making process is as follows: vehicles that are able to drive faster than 6 km/h (3.7 mph) and would be driven on public roads must fulfill national requirements or, if possible, apply for a European Economic Community (EEC)-type approval. If the vehicle manufacturer applies for an EEC-type approval, it also must prove that it has introduced a quality control system to assure the conformity of production (COP). A typical application field is Germany. Table 4.1 shows the rules for cars with up to eight seats for Europe EEC, Germany, and the Economic Commission for Europe for accident avoidance and the minimization of injuries. Table 4.2 shows the relevant rules in the United States [4-1].

Based on the agreement of 1958 with respect to common rules for the approval of parts and vehicles, the United Nations Economic Commission for Europe is working to harmonize the different rules worldwide. With the adjustment of October 16, 1995, the working title now has the following meaning: "New Agreement concerning the adoption of uniform technical prescriptions for wheeled vehicles, equipment, and parts that are used in road vehicles..."

TABLE 4.1
REQUIREMENTS FOR ACTIVE SAFETY IN EUROPE

	EC Directive	ECE Regulation	StVZO*
Requirements for Active Vehicle Safety (Accident Prevention):			
Steering equipment	70/311/EC	R 79	§ 38
Brake systems	71/320/EC	R 13	§ 41
Replacement brake pads/shoes	71/320/EC	R 90	§ 22
Equipment for acoustic signals	70/388/EC	R 28	§ 55
Field of vision	77/649/EC	R -	§ 35b
Defrosting and defogging systems for glazing	78/317/EC	R -	§ 35b
Windshield wipers and washers	78/318/EC	R -	§ 40
Rearview mirrors	71/127/EC	R 46	§ 56
Heaters (engine waste heat)	78/548/EC	R -	§ 35c
Gas heaters, auxiliary heaters	—	R -	§ 22a
Installation of lighting and lighting-signaling devices	76/756/EC	R 48	§ 49a, 53a
Reflex reflectors	76/757/EC	R 3	§ 53
Clearance lamps, taillamps, stop lamps	76/758/EC	R 7	§ 51, 51b, 53
Side marker lamps	76/758/EC	R 91	§ 51a
Turn signal lamps	76/759/EC	R 6	§ 54
Headlamps for high beam and/or low beam	76/761/EC	R 1, 8, 20 R37	§ 50
and their light sources	—	—	§ 22a
Gaseous discharge headlamps and their light sources	— —	R 98 R 99	— —
Front fog lamps	76/762/EC	R 19	§ 52
Rear fog light (fog taillamp)	77/538/EC	R 38	§ 53d
Backup lamps (reversing lamps)	77/539/EC	R 23	§ 52a
Parking lamps	77/540/EC	R 77	§ 51c
Rear registration plate illumination devices	76/760/EC	R 4	§ 60
Reverse gear and speedometer equipment	75/443/EC	R 39	§ 39, 57
Interior equipment (symbols, warning lights)	78/316/EC	R -	§ 30
Wheel covers	78/549/EC	R -	§ 36a
Tire tread depth	89/459/EC	R -	§ 36
Tires and tire mounting	92/23/EC	R 30	§ 36
Towing capacity, hitch vertical load	92/21/EC	R -	§ 42, 44
Towing equipment (trailer hitches)	94/20/EC	R 55	§ 43
Pedal arrangement	—	R 35	§ 30

*(Straßenverkehrs-Zulassungsordnung = Road Traffic Licensing Regulation)

TABLE 4.1
REQUIREMENTS FOR ACTIVE SAFETY IN EUROPE *(cont.)*

Requirements for Passive Vehicle Safety (Mitigation of Injuries):

Interior fittings (protruding elements)	74/60/EC	R 21	§ 30
Steering mechanism (behavior in an impact)	74/297/EC	R 12	§ 38
Frontal impact, occupant protection	96/79/EC	R 94	—
Side impact, occupant protection	96/27/EC	R 95	—
Seat-belt anchorages	76/115/EC	R 14	§ 35a
Seat belts and restraint systems	77/541/EC	R 16	§ 35a
Seats, seat anchorages, head restraints	74/408/EC	R 17, 25	§ 35a
Head restraints	78/932/EC	R 17, 25	§ 35ac
External protection	74/483/EC	R 26	§ 30c
Fuel tanks and rear underride protection	70/221/EC	R 58	§ 47, 47c
Liquefied petroleum fuel systems	—	R 67	§ 41a, 45, 47
Doors (locks and hinges)	70/387/EC	R 11	§ 35e
Front and rear bumpers	—	R 42	—
Rear-end collisions (not applicable to Germany)	—	R 32	—
Child restraint systems	—	R 44	§ 22a
Safety glazing	92/22/EC	R 43	§ 40
Electric propulsion (safety)	—	R 100	§ 62

TABLE 4.2
RELEVANT RULES IN THE UNITED STATES [4.1]

FMVSS	Contents
101	Controls and displays
102	Transmission shift lever sequence, starter interlock, and transmission braking effect
103	Windshield defrosting and defogging systems
104	Windshield wiping and washing systems
105	Hydraulic and electric brake systems
106	Brake hoses
108	Lamps, reflective devices, and associated equipment
109	New pneumatic tires
110	Tire selection and rims
111	Rearview mirrors

TABLE 4.2
RELEVANT RULES IN THE UNITED STATES [4.1] *(cont.)*

FMVSS	Contents
113	Hood latch system
114	Theft protection
116	Motor vehicle brake fluids
118	Power-operated window, partition, and roof panel systems
119	New pneumatic tires for vehicles other than passenger cars
120	Tire selection and rims for motor vehicles other than passenger cars
121	Air brake systems
124	Accelerator control systems
129	New non-pneumatic tires for passenger cars
135	Passenger car brake system
201	Occupant protection in interior impact
202	Head restraints
203	Impact protection for the driver from the steering control system
204	Steering control rearward displacement
205	Glazing materials
206	Door locks and door retention components
207	Seating systems
208	Occupant crash protection
209	Seat-belt assemblies
210	Seat-belt assembly anchorages
212	Windshield mounting
213	Child restraint systems
214	Side impact protection
216	Roof crush resistance
219	Windshield zone intrusion
301	Fuel system integrity
302	Flammability of interior materials
303	Fuel system integrity of compressed natural gas vehicles
304	Compressed natural gas fuel container integrity

Application for membership is on a voluntary basis. This means that countries that are members of the United Nations but not of the EEC could join this organization.

Since March 1998 and November 1999, the EEC and Japan, repsectively, became members of this commission. More than 100 rules are in effect, and some of those rules could be used as part of the EEC-type approval. Additional actions to minimize trade barriers among the continents of the United States, Japan, and Europe started with the Transatlantic Economy Dialog. The goal is to improve trade relations between Europe and the United States and to reduce trade barriers. A further commitment was given to create, on a more global basis, rules for vehicles. In the International Harmonized Research Activities (IHRA), the following themes, led by the countries below, are handled:

Australia:	Side impacts
Italy:	Frontal impacts
United Kingdom:	Vehicle compatibility
United States:	Biomechanics
Canada:	Intelligent transportation systems
Japan:	Pedestrian safety

Guided by the IHRA Steering Committee, the number of harmonized rules should be increased. As of February 2000, approximately 80% of the rules were capable of being harmonized, if the car manufacturer accepts the higher level of safety in the other continents.

4.1 Reference

4-1. Compiled from sources of National Highway Traffic Safety Administration (NHTSA) and Federal Motor Vehicle Safety Standards (FMVSS), Washington, DC, United States.

5.

Accident Data

This chapter provides some general data related to accident statistics. In addition to this general data, we can say that for more complex engineering work on vehicle safety, very detailed information is necessary. Worldwide, there are many different organizations actively involved in accident research work, such as the governments, universities, insurance associations, and vehicle manufacturers. In direct meetings among the experts and in many conferences, the facts and conclusions are discussed.

Figure 5.1 shows the development of traffic fatalities as a function of billion vehicle kilometers for various countries from 1970 to 1996.

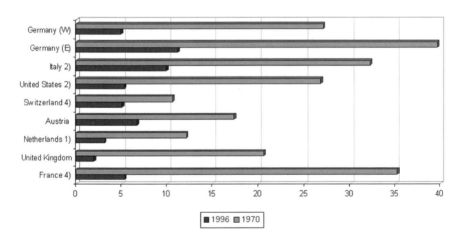

Figure 5.1 Development of the fatality rate in road traffic in Western Europe and the United States (fatalities per billion vehicle kilometers). (Source: Ref. 5-1.)

Although the rate of improvement is impressive, too many accidents continue to occur, with severe and fatal injuries on public roads. For example, in 2000, Europe had approximately 35,000 fatalities per year with car involvement on roads. If we look more closely at the distribution of fatal accidents, Figure 5.2 shows the various participants in Germany as a function of years. The group of passenger car occupants is the largest.

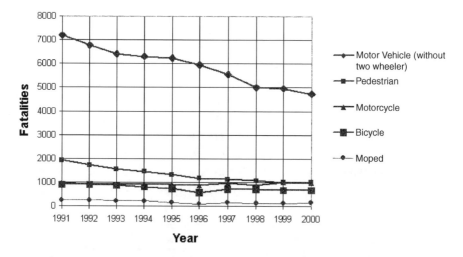

Figure 5.2 Number of deaths in accordance with traffic participants in Germany. (Source: Ref. 5-2.)

Another strong influence is the road itself. Figure 5.3 demonstrates the differences in the types of roads in Germany.

This information is relevant in absolute numbers and also if you compare the fatalities per kilometers driven where, for example, the safety-per-kilometers driven is approximately a factor of four better on highways than on other roads. There is one special effect in Germany and in other European countries for country roads. Approximately 30% of all fatalities on country roads occurred because of a large concentration of impacts with trees. How this number will be reduced as a result of the installation of electronic stabilization control and head side airbags will be seen in the future.

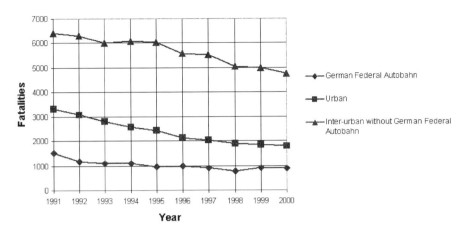

Figure 5.3 Number of deaths as a function of road type.
(Source: Ref. 5-2.)

Figure 5.4 shows an international comparison with respect to the reduction of traffic deaths per billions of kilometers driven on highways in various countries.

There also is no clear indication that countries with speed limits on highways have a lower accident rate than others, in terms of fatal accidents. If we examine the type of crash for car occupants, we also find interesting results, as shown in Figure 5.5 [5-4, 5-5].

In the United States, 40% of all crashes are single-car accidents. This also is true for Germany. However, we find differences in areas such as cars to pickup trucks and sport utility vehicles, pedestrians, and two-wheel vehicle accidents. It also is important to look at the driver, who often is the main cause of the accident. In addition to the more direct parameter, which is related to the specific items discussed in Chapter 6 on accident avoidance, two general observations could be made. There definitely exists a correlation between age and risk. Figure 5.6 shows the results of a study conducted in the United States.

From Figure 5.6, we can conclude that young drivers have a high risk for causing crashes. This higher risk will be increased by the number of passengers in

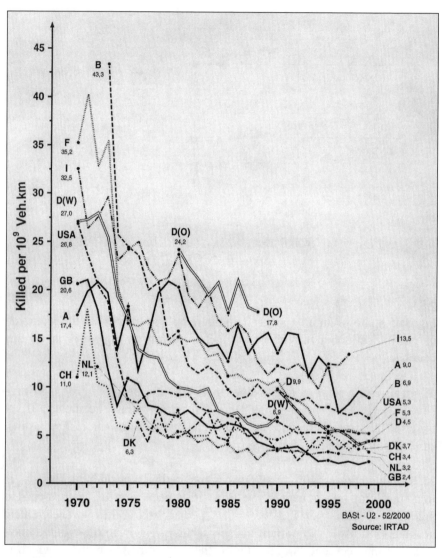

*Figure 5.4 Death rate on highways—an international comparison.
(Source: Ref. 5-3.)*

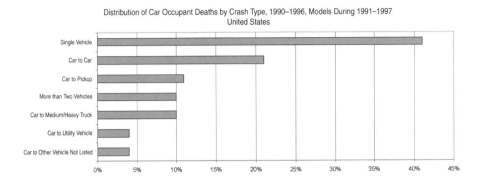

Distribution of Car Occupant Deaths by Crash Type, 1990–1996, Models During 1991–1997
United States

Distribution of Passenger Vehicle Accidents, Germany, 2000

*Figure 5.5(a) Distribution of accident types in the United States.
(b) Distribution of accident types in Germany. (Source: Ref. 5-4.)*

21

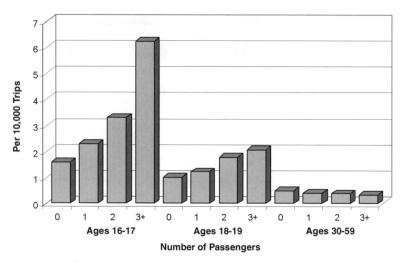

Figure 5.6 Crash rates by driver age and passenger presence.
(Source: Ref. 5-6.)

the car [5-6]. A study conducted in Germany, as shown in Figure 5.7, compares the number of pedestrians and car occupants injured in accidents.

Figure 5.7 shows a high risk in both categories, that is, accidents per kilometers driven and accidents per time for children and young people [1-1]. Another question, which often is asked in private discussions, is whether women or men have more or fewer accidents. As illustrated in Figure 5.8, there seems to be a general advantage for women with regard to a lower number of accidents, especially in single-car crashes.

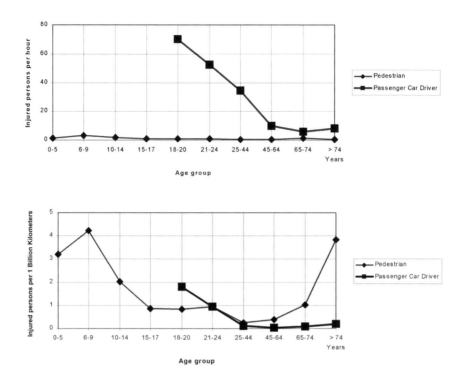

Figure 5.7 Accident rates for pedestrians and passenger car drivers in Germany. (Source: Ref. 1-1.)

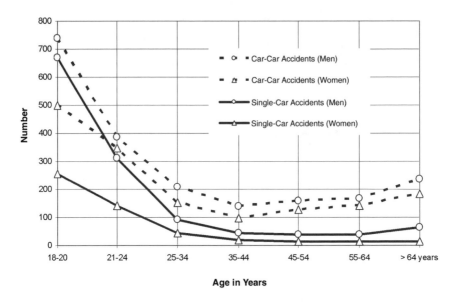

Figure 5.8 Passenger car accidents per 1,000 owners of drivers licenses and one million kilometers driven. (Source: Ref. 5-7.)

5.1 References

5-1. Bundesanstalt für Straßenwesen, BAST, unpublished data, Bergisch Gladbach, Germany.

5-2. Friedel, Bernd, unpublished data, BAST, Bergisch Gladbach, Germany.

5-3. Brühning, E. et al. "Entwicklung der Verkehrssicherheit auf europäischen Autobahnen," *Straße und Autobahn*, 47 (1996), No. 1, pp. 22–26.

5-4. *Verkehr in Zahlen 2000*, Deutscher Verkehrsverlag, Bundesministerium für Verkehr (Ed.), Hamburg 2001, ISBN 3-87154-259-9.

5-5. Insurance Institute for Highway Safety, Special Issue: "Vehicle Compatibility in Crashes," October 1999, Arlington, VA, United States.

5-6.   Insurance Institute for Highway Safety annual status report, Vol. 34, February 1999, Arlington, VA, United States.

5-7.   *Auto Motor Sport*, 21/1999, Stuttgart, Germany.

6.

Accident Avoidance

In the category of accident avoidance are all measures that help in a positive manner to prevent accidents. Related to the vehicle driver, we can state that all measures that contribute to an easier, safer, and more comfortable drive also help to avoid accidents. Driver assistance systems should support the driving function.

In addition to the technical features of drive-by-wire, it is very important to the man-machine interface to define precisely the responsibilities that the driver should retain. The responsibility of the driver also is one basic requirement of the Wiener world agreement for "Driver Assistance Systems." It is easy to understand that this does not have to be a rigid definition. Technological progress and a growing understanding of the man-machine interface will allow more continuous step-by-step progress in this area. One fundamental rule must be fulfilled: New technical features should not be installed in any case if those features have a foreseeable negative effect, even if this effect is very small.

6.1 Human Factors

Humans as traffic participants play a major role with respect to failure that might lead to an accident. Table 6.1 shows the involvement of various groups related to the occurrence of accidents.

The main group is car occupants. Some data indicate that the driver is responsible for more than 80% of all accidents. If we take an average of all German states, we find the data in the police reports for 1999 with respect to the group that is responsible for these accidents, as shown in Table 6.2 [6-1].

TABLE 6.1
DISTRIBUTION OF TRAFFIC PARTICIPANT GROUPS
WITH RESPECT TO ACCIDENTS

Type of Traffic Participant	Europe [%]	Germany [%]	United States [%]
Passenger car occupant	42–63	51	73.5
Motorized two-wheeler	14–27	14	7.5
Bicycle driver	5–20	9	2.0
Pedestrian	14–34	23	14.5

TABLE 6.2
THE ORIGINATOR OF ACCIDENTS WITH INJURIES,
AS A FUNCTION OF TRAFFIC PARTICIPANTS

Responsible for the Accident	%
Driver (excluding cyclists)	76.6
Cyclist	10.4
Pedestrian	5.3
Road conditions	4.5
Technical failures on cars	0.84
Others	2.36

Although the driver's group is the most significant as the major cause of accidents, we must be careful not to neglect other influences. The driver is influenced by his or her health and physical condition, qualifications and experience, driving education, ability for orientation, quality of the man-machine interface, climate, and comfort related to the vehicle. In more detailed studies, we also find different figures for the driver being responsible for causing the accident. For example, a vehicle with accident prevention measures or higher comfort levels also might lead to a lower accident rate. If we take this into consideration as a result of discussion with various sources, we find approximately 65% human-based failures, 30% environmental, and 5% vehicle-related. If we examine a detailed analysis of the single failures for the driver, the general distribution for the 1999 statistic in Germany [6-1] could be taken from Table 6.3.

TABLE 6.3
CAUSES OF ACCIDENTS FOR ALL TYPES OF VEHICLES, INCLUDING MOTORCYCLES, BIKES, AND OTHER VEHICLES

Speed too high	18.6%
Right of way (intersections)	13.4%
Turns, drive in and out, turn around	13.9%
Driving too close to another car	12.3%
Influence of alcohol	5.4%
Misuse of the road	6.7%
Failure during passing	5.8%
Wrong behavior against pedestrians	5.2%
Others	18.7%

The mental and physical condition of a driver is one of the most important factors to consider when discussing the avoidance of accidents. Mistakes made by the driver have varied causes. Often, he or she dismisses the possibility of an accident, a momentary over-estimation of self, lack of driving experience, insufficient physical condition, and impairment due to the consumption of alcohol or drugs. Alcohol consumption affects the safe driving ability of the driver.

Figure 6.1 shows an investigation with various people and blood alcohol content (BAC), driving a specific course on the Volkswagen driving simulator. The test persons had to drive a certain distance, where the number of mistakes as a function of BAC was measured. A significant deviation from normal driving was observed with an alcohol content of approximately 0.6%. In a study of accidents, the increase in the risk to cause an accident as a function of BAC is as shown in Table 6.4.

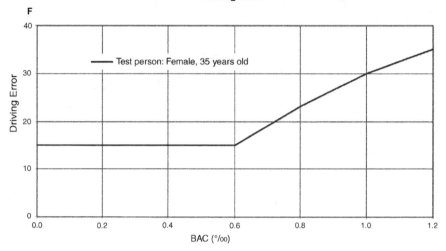

Figure 6.1 *Driving failures as a function of blood alcohol content (BAC).*
(Source: Ref. 1-1.)

TABLE 6.4
RISK INDEX OF ALL ACCIDENTS
AS A FUNCTION OF BLOOD
ALCOHOL CONTENT (BAC) [6-2]

BAC Level	Risk Index K
0	1
0.01–0.3	1.5
0.31–0.5	2.5
0.51–0.8	5
0.81–1.3	20
1.31–......	25

In most countries, we find legislation limits on blood alcohol content as defined by the authorities. From the first work on experimental safety vehicles, there have been many investigations on keeping the drunk driver from being able to start the vehicle. All invented devices until now have failed. Either these devices were so complicated that even drivers without any blood alcohol content were unable to start the car, or drunk drivers overcame the devices with the help of a "solver" person. In addition to this negative influence of alcohol, we find drugs as a cause of accidents, although the exact figures are unknown.

Another element of research is that tired drivers may fall asleep while driving the vehicle. Drowsiness sensors might be one solution for these critical situations. For example, systems would watch the movements of the eyelid via cameras and give an acoustic or mechanical (vibration) warning signal. Other possibilities would be to monitor the steering wheel movement or the pulse frequency and breathing of the driver.

In general, we can state that for future work on vehicle safety, detailed knowledge about the behavior of the driver and the other traffic participants from the results of a more detailed accident investigation becomes increasingly important. However, for new systems, only a theoretical prediction of the effect can be made.

Some single elements of accident avoidance measures are described in the following sections.

6.2 Comfort and Ergonomics

To a certain degree, there are direct relationships between comfort and vehicle safety. All aspects that contribute to a more comfortable drive also help to reduce accidents. This includes comfortable entry and egress from the vehicle, which has the entrance and seat not too low above the road, a good adjustable seat, and a seat back with lumbar support. Several other elements, which vehicle occupants today take for granted, contribute to an increased level of comfort. These include seat-belt latches on the seat, height adjustments of the upper anchorage point, and vertically and horizontally adjustable steering wheels. Figure 6.2 illustrates a connection between the comfort of the seat and driver fitness.

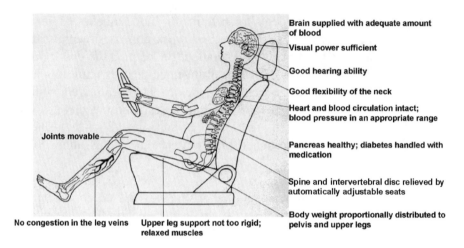

Brain supplied with adequate amount of blood

Visual power sufficient

Good hearing ability

Good flexibility of the neck

Heart and blood circulation intact; blood pressure in an appropriate range

Pancreas healthy; diabetes handled with medication

Spine and intervertebral disc relieved by automatically adjustable seats

Body weight proportionally distributed to pelvis and upper legs

Joints movable

No congestion in the leg veins

Upper leg support not too rigid; relaxed muscles

Figure 6.2 Accident avoidance measures. (Source: Ref. 1-1.)

Other minor items can have a great positive influence on driver comfort. These include electrically adjustable and heated mirrors, power windows, power steering, heated wiper systems, rain sensors, and automatically adjustable interior rearview mirrors to avoid blinding from glare. Although fuel consumption due to the higher amount of electrical devices in the car often increases slightly, the benefit for safety justifies these installations. One positive example is the much higher installation rate of air conditioning systems in cars. For many years, this was the domain of cars in the United States, Japan, and the upper-class segment in Europe. Meanwhile, in many countries, air conditioning in cars has become standard equipment. As shown in Figure 6.3, from an investigation performed by the company Behr [6-3], we can find a direct correlation between the accident rate and the heat loading on the driver.

The weather likewise directly influences the occurrence of accidents. On wet roads, we find a 20% increase in accidents, with a high thermal load slightly above a 20% increase.

Another comfort item is low interior noise. Figure 6.4 shows the dB (A) level of an optimization process in a vehicle body as a function of engine rpm.

% change in average number of accidents

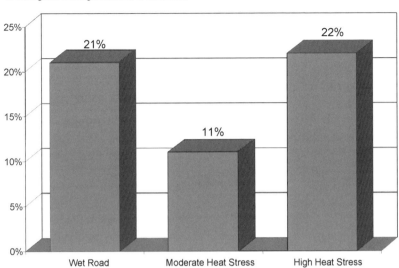

Increase in number of in-town accidents in case of wet roads and heat stress of the driver.

Figure 6.3 Influence of weather conditions and heat loading of the driver.

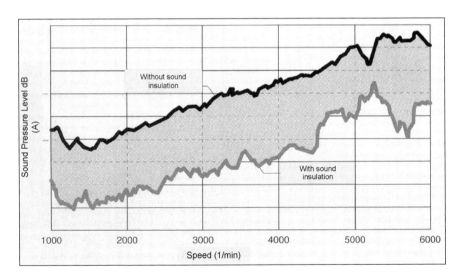

*Figure 6.4 Reduction of the total noise pressure level due to
isolation measures. (Source: Ref. 6-4.)*

In this connection, controlled measures for noise and vibration reduction are becoming necessary to eliminate special problems. Not only is a low level of noise important, but also the fact that single spikes should be avoided. In addition to a low level of interior noise, an optimized climate control in the passenger compartment, with the possibility of guiding the airflow to individual parts of the human body (and on high-class cars, the ability of the driver and passenger to adjust the climate control separately) is standard equipment. The vertical oscillations of the vehicle also influence the comfort level of the driver and the vehicle occupants. Figure 6.5 demonstrates the relationship between oscillation and damping comfort and contact between tire and road. The optimal layout is within the limited values (e.g., below a K-factor of 10).

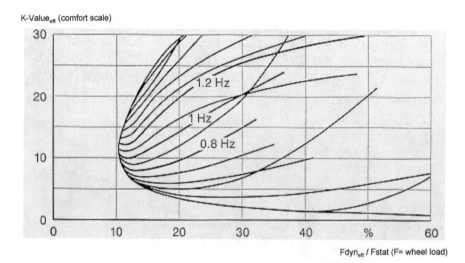

Figure 6.5 K-scale as function of wheel load and vehicle natural frequency. (Source: Ref. 6-5.)

According to Richter [6-5], a good compromise for the layout of the chassis is shown, if the data as shown are fulfilled, although the necessary low unsprung mass could not always be achieved as easily. Engineering research and development activities today are leaning toward the development of the active chassis, which also reduces the critical oscillation frequencies. Figure 6.6 shows the design of an active chassis control.

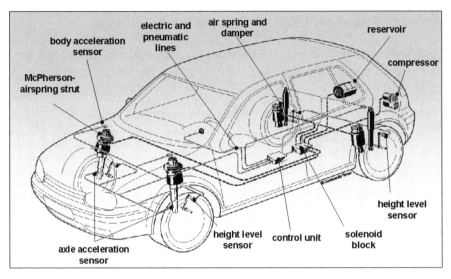

*Figure 6.6 System architecture for an electronic suspension system.
(Source: Ref. 6-6.)*

The ergonomic design for the working-place "driver" and for the man-machine interface is evident with regard to accident avoidance capability. In the BMW 700 series for the calendar year 2001, the redesign of the operating concept tries to integrate the following features of the intuitive and adaptive man-machine interface [6-7]:

- Intuitivism
- Transparency
- Failure robustness
- Effectiveness
- Individuality
- Safety
- Learning ability
- Flexibility
- Adaptive
- Socially compatible
- Multi-modularity

The reaction of the market today favors the new solution.

The working-place driver also must be designed for different sizes of humans. The two-dimensional design dummies cover the 5% female up to the 95% male, as shown in Figure 6.7.

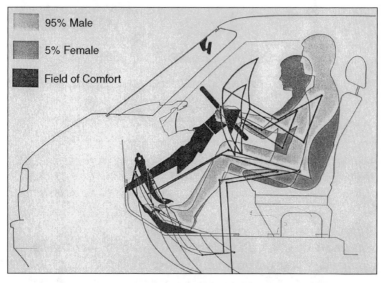

95% Male

5% Female

Field of Comfort

Figure 6.7 Steering wheel and pedal layout in a microbus.
(Source: Ref. 1-1.)

From studies at the University of Kiel (Germany), we know that an average man of the young generation (i.e., 20–25 years old) has a total body length that is approximately 12 cm (4.7 in.) longer compared with the group of people between the ages of 16 to 60. For the actual layout of the interior of the vehicle, car manufacturers meanwhile are using more realistic data. Because of a large number of legal requirements, a three-dimensional test device is used, as shown in Figure 6.8 [6-8].

With this device, the seating reference point can be determined. The seating reference point is the basis for the field of view and for the seat-belt anchorage point location. In addition, designers use various mathematical models.

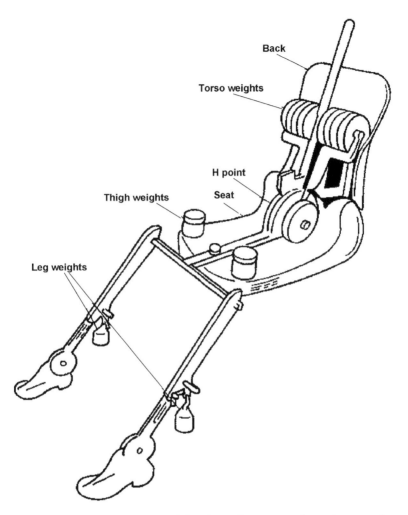

Figure 6.8 Description of the three-dimensional machine tools.

Computer technology today allows a much better simulation of the ergonomic design of the working-place driver. In the virtual product creation process, different sizes and three-dimensional effects can be simulated. Figure 6.9 demonstrates Ramsis, a three-dimensional computer model simulating the driver [6-8].

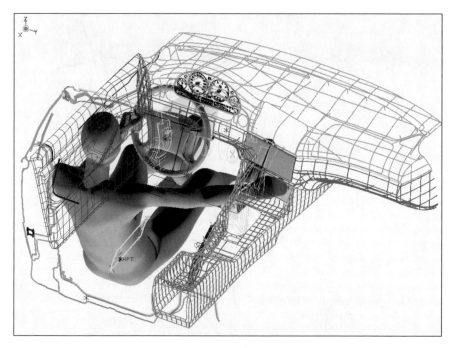

Figure 6.9 Instrument panel layout by Ramsis.

For the layout of the interior of the vehicle, not only are the driver's physical dimensions considered, but the interior lighting systems are receiving more attention. During the entry and egress process of vehicle occupants, lighting should contribute to safety and orientation. During driving, the driver should not be disturbed by the reading light or the instrument panel lights. The instrument panel should be read easily by the driver and have a color that is optimal for most drivers. The instrumentation should not produce reflected glare in the windshield. Although head-up displays give the designer an additional free parameter, these have not achieved a large market penetration. This might change if more warning functions are given via head-up displays to the driver. A first solution is already on the market.

In terms of the driver's field of view, numerous safety requirements exist. The driver should have a good all-around view, related to the vehicle itself, and a free field of view through the windshield. The requirements in the field of view and the wiped pattern on the windshield are defined in FMVSS 103 and 104 [6-9, 6-10]. The field of view is divided into Fields A, B, and C. Figure 6.10 shows these areas for a compact car.

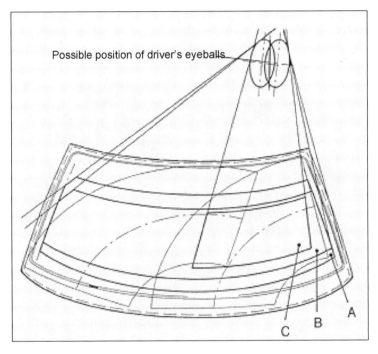

Figure 6.10 The field of view for a compact passenger car.

The areas in the field of view are defined by tangential surfaces on the ellipse of the driver's eyeballs. The defined areas must be wiped with a specified percentage (more than 80% for Field A, more than 94% for Field B, and more than 99% for Field C). Similar requirements must be fulfilled in other countries such as Europe. The determined areas also serve for the checking of the performance of the windshield heating system (defroster/defogger). The vehicle under investigation is tested in accordance with test procedure SAE J902 [6-11]. The vehicle is pre-conditioned at an environment temperature of –18°C (–0.4°F). In a period of less than 40 min, Field A must be defrosted by more than 80% and Field C by 100%.

The all-around view includes the exterior and interior mirrors. In the area of optimizing the field of view, we also find new wiper systems that by their geometric layout minimize the unwiped area. Several support systems also are available in some cars. These include heated wiper nozzles and heated outside mirrors, electrical mirror adjustment, electrically heated windshields, rain sensors (see Figure 6.11) that automatically switch on the windshield wipers, and automated interior anti-glare rearview mirrors.

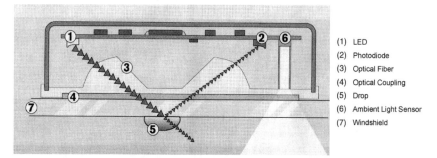

(1) LED
(2) Photodiode
(3) Optical Fiber
(4) Optical Coupling
(5) Drop
(6) Ambient Light Sensor
(7) Windshield

Figure 6.11 Layout of a rain sensor. (Source: Ref. 6-12.)

A more advanced system uses sensors that are installed in the outside mirrors to observe both sides of the car, to prevent an accident in a critical lane-changing maneuver, or to prevent a possible accident with a pedestrian or cyclist in a right-turn situation. It has not yet been determined how the critical situation should be transmitted to the driver. Several solutions are possible, ranging from an acoustical warning signal to an increase of steering wheel torque, as shown in Figure 6.12.

Other systems that indirectly improve the field of view are obstacle identification systems that measure at low-speed driving the distance to other obstacles and provide an acoustical or light signal to the driver, as shown in Figure 6.13. The following functions could be fulfilled [6-13]:

• View of blind spot

• Parking assistant

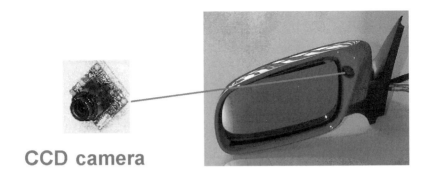

CCD camera

Figure 6.12 Rearview mirror with integrated camera electronics for detecting objects in vehicle blind spots. (Source: Ref. 6-13.)

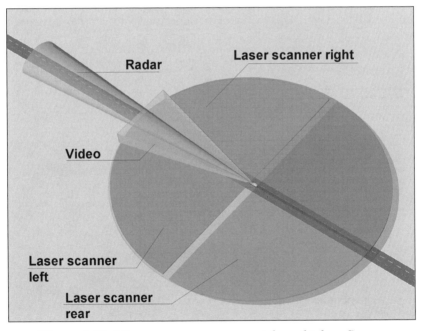

Figure 6.13 Microwave sensor concept for vehicles. Sensor, video sensor, and laser scanner. (Source: Ref. 6-13.)

- Adaptive cruise control (ACC)—Short-range sensor for stop-and-go function

- Pre-crash detection

The connection to vehicle safety is given by the fact that small amounts of damage to vehicle exterior components and therefore to the lights and signals are prevented.

One example of a large number of research activities to improve the driver's all-around view is demonstrated in Figure 6.14 by the Bosch company. If we analyze Figure 6.14 closely, we could imagine that the large number of sensors are able to improve the safety of the vehicle in several areas if the man-machine interface is designed correctly. The driver assistant systems can support several functions, such as the parking pilot, with ultrasonic sensors in the bumper system, up to 1.5 m (5 ft); adaptive cruise control (ACC); a long-range radar sensor of 77 GHz; and a short-range sensor of 24 GHz for identifying obstacles close to the vehicle. In the future, cameras will be used in addition to sensors. This provides an opportunity not only to measure the

Protection zones

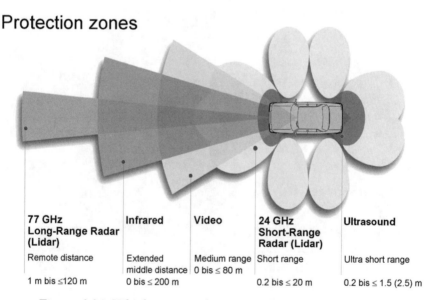

77 GHz Long-Range Radar (Lidar)	Infrared	Video	24 GHz Short-Range Radar (Lidar)	Ultrasound
Remote distance	Extended middle distance	Medium range 0 bis ≤ 80 m	Short range	Ultra short range
1 m bis ≤120 m	0 bis ≤ 200 m		0.2 bis ≤ 20 m	0.2 bis ≤ 1.5 (2.5) m

Figure 6.14 Vehicle surround sensors. (Source: Ref. 6-14.)

distance to objects in front of the vehicle but to identify of the types of objects. Other sensors might provide information about road and bad weather conditions such as snow and fog.

In relation to the area of field of view, we also must add the function of the signals and the lights of vehicles. The signals are not used to improve vision but to identify the vehicle and to show the purpose of the driver's action, if he or she changes the direction of the drive path or applies the brakes. For passenger cars, a third brake light at an increased height above the ground is installed. This light allows drivers in the vehicle behind the braking vehicle to more quickly recognize a braking situation, thereby helping to speed up their reaction time to apply the brakes. There are also some discussions regarding whether the third brake light should change its color intensity in relation to braking force.

With respect to the development of headlamps, Figure 6.15 shows the changes as a function of years.

Related to the design of headlamps, the trend to install the technical elements behind a glass or plastic cover is continuing. The shape of the headlamp cover is designed for low aerodynamic resistance and pedestrian protection.

For the layout of headlamps in terms of lighting intensity, we must make sure that the headlamps provide an optimized field of view for the driver, without blinding other drivers on the road. In general, we can ask for the fulfillment of the following requirements:

• The road should be illuminated symmetrically and without spots.

• The driver should have a good view of the road shoulder on his side.

• The majority of light should cover the driving path.

• Scatter width should be sufficient.

• Avoid blinding the oncoming traffic.

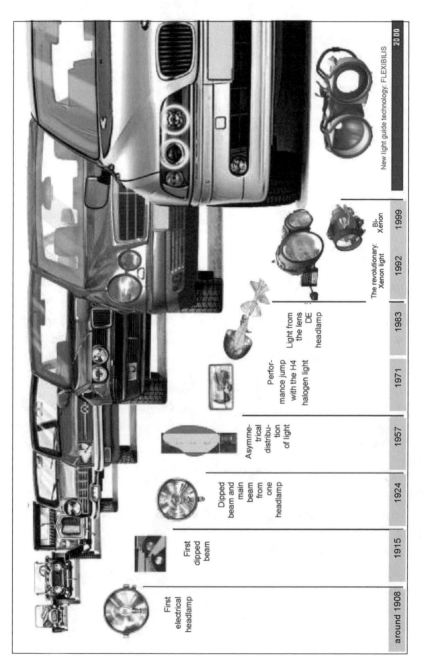

Figure 6.15 Product development of headlamps. (Source: Ref. 6-15.)

- The driving path in front of the vehicle up to a distance of 40 m (131 ft) should not be illuminated to an extreme level because of the possibility of blinding from a wet road, due to the reflection of light from oncoming traffic.

Also in the lighting equipment, we find some driver assistance systems. For example, the automatic vertical adjustment of headlamps is especially requested for xenon light. The automatic headlamp leveling device must be able to identify the position of the vehicle relative to the road. This is accomplished by use of angular sensors. In the future, even more advanced systems are foreseeable. One example is additional infrared headlamps to identify pedestrians or other road obstacles in darkness.

In addition to vertical adjustment, a computer-controlled horizontal headlamp adjustment based on either the steering wheel angle or, in the future, in combination with a positioning sensor such as GPS (global positioning satellite) is possible. The benefit of this solution lies in the fact that road curves are illuminated before the vehicle has reached the next curve of the road. The adaptive front lighting will have several dynamic functions. One function is the town light, which illuminates a larger area compared to the asymmetric beam used today. The other function is the country light, which will include the recognition of road course obstacles and road signs in a more efficient manner, as shown in Figure 6.16 [6-16].

6.3 Acceleration and Braking

Because of the big difference among passenger cars with respect to engine power and vehicle mass, and because of the necessity to keep traffic flowing, establishment of minimum performance with respect to power and torque is necessary. This is a request not only for Europe but for North America and some states in Asia. For most passenger cars, front-wheel drive is standard. To achieve sufficient traction and brake capability, many additional features are available in production cars. For traction control, we find that different names for the system are used by the car producer. In principle, this system uses a portion of an antilock brake system (ABS) for electronic differential control. This device is sufficient to prevent a traction wheel from spinning when starting a vehicle on a surface with a different coefficient of friction. At higher speed (e.g., above 20 km/h [12 mph]), the effect is reduced. Above

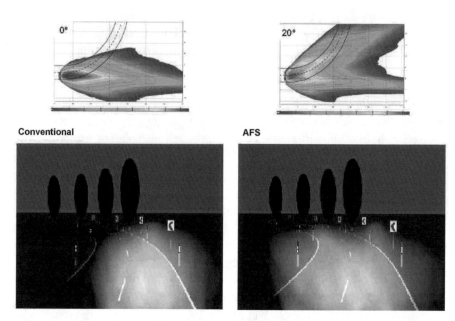

Figure 6.16 Advanced frontlight system (AFS) (dynamic curve light).
(Source: Ref. 6-16.)

40 km/h (25 mph), the system is switched off to avoid a negative influence
with respect to safe driving under high speed. Figure 6.17 shows the general
function of such a system.

With other systems (e.g., automatic stability control plus traction [ASC+T]),
during a wheel spin of one drive wheel, the propulsion force via an electronic
engine management system is reduced. In addition, through a brake inter-
vention at the spinning wheel up to 40 km/h (25 mph), this effect is again
reinforced. The electronic differential system (EDS), propulsion slip control
(ASR), and ASC+T systems are sufficient for many driving conditions. A
more sophisticated device to control the acceleration capability is four-wheel
drive. Several four-wheel-drive systems are on the market. Figure 6.18 is an
overview of the different driveline systems, including four-wheel drive.

Four-wheel drive shows positive function not only under acceleration and
complicated road conditions but during hydroplaning as well. A pioneer for

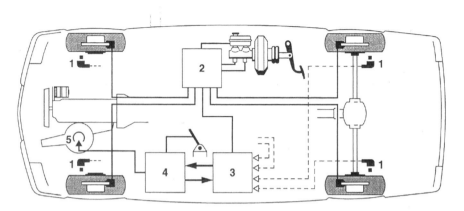

Figure 6.17 An antilock brake system/propulsion slip control system (ABS/ ASR [Acceleration Slip Regulation]) for passenger cars. (1) Speed sensor; (2) ABS/ASR hydraulic unit; (3) ABS/ASR ECU; (4) Electronic throttle ECU; (5) Throttle gasoline engine. (Source: Ref. 6-17.)

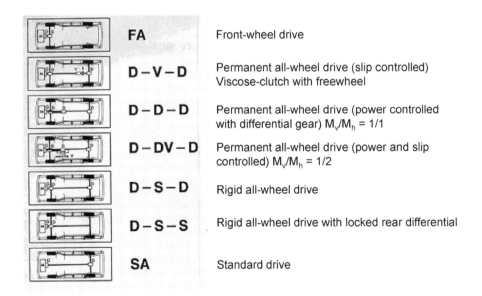

Figure 6.18 View of driveline systems. (Source: Ref. 1-1.)

permanent four-wheel drive for passenger cars was Audi AG. Volkswagen's newest system is based on a hydraulic electronically controlled clutch, where the front and rear axle are connected, if this is required by unusual road situations, as shown in Figure 6.19.

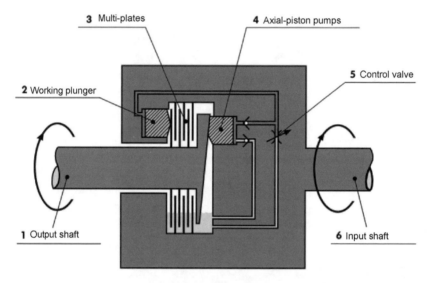

Figure 6.19 Function of the four-wheel-drive Haldex clutch.
(Source: Ref. 6-18.)

For four-wheel drive cars, which are used primarily on normal roads, the four-wheel system is disconnected to avoid vehicle spinning, if, during down-hill driving, the brakes are applied.

For most owners of vehicles, the vehicles with EDS/ASR systems are suffi-cient. This is why four-wheel drive remains a niche market, although high-powered vehicles and sport utility vehicles (SUVs) have a higher and increasing installation rate of four-wheel drivelines in the vehicles.

6.3.1 Adaptive Cruise Control (ACC)

Adaptive cruise control (ACC) is one system that is already in series production. It started in connection with upper-class models and now is available in the compact class as well. Adaptive cruise control identifies the vehicle in front of the ACC-equipped vehicle, determines its position and speed, and maintains (through braking and acceleration) the desired distance between the two vehicles. The main benefit is achieved on roads with high speed limits, such as highways and the Autobahn in Germany. To maintain the desired speed and safe distance from the vehicle driving in front, the car equipped with ACC uses a 77-GHz radar sensor to identify moving objects to a distance of 100 m (322 ft) in front of the vehicle and at an angle of $\pm 4°$ to the centerline. With the determined data, the electronic control unit calculates the relative distance to the moving object in front of the ACC-equipped vehicle. If no other vehicle is in the front of the ACC-equipped vehicle, the vehicle accelerates to the selected speed. Although the driver receives strong support for this drive function, he or she remains responsible for control of the vehicle. Figure 6.20 shows some parts of the adaptive ACC by Bosch (the radar and the electronic control unit).

Technical Data

Temperature range	-40 to 80°C
Power consumption	13 W
Package size:	12.4 cm H
	9.1 cm W
	9.7 cm D
Weight	<0.6 kg

Bracket for vehicle-specific
   mounting including alignment
   mechanics

Figure 6.20 Components of adaptive cruise control (ACC).
(Source: Ref. 6-19.)

6.3.2 Brakes

Standard equipment on passenger cars today is a hydraulic two-circuit brake system with two independent working brake circuits. A typical layout is the diagonal configuration of the brake (one for a front and a rear wheel). Chassis design with a negative kingpin offset supports directional stability during the braking maneuver. A load- or pressure-dependent braking force regulator prevents locking of the brakes at the rear axle. This guarantees stability during braking.

Figure 6.21 shows the importance of a short reaction time and a rapid increase in brake pressure for the prevention of accidents.

Some production cars use a so-called "brake assistant." In emergency braking situations, due to a pressure boost control or via a dual rate booster with an emergency valve, this brake assistant rapidly increases brake pressure, even at relatively low brake-pedal force. The design of the brake assistant can be described as follows:

The signal of the pressure sensor determines from the pressure level and pressure gradient if an emergency braking situation exists. This increases

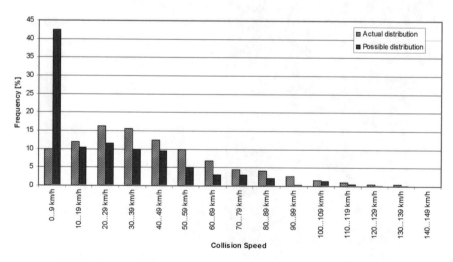

*Figure 6.21 Distribution of accidents, as a function of collision speed.
(Source: Ref. 6-20.)*

the braking force, if necessary, in an initial phase. If the pressure applied by the driver is reduced, the brake pressure is increased automatically again. The brake pressure also is increased if the driver is not pushing the brake pedal hard enough. Figure 6.22 shows details of the control function [6-21]. The brake assistant in a not-too-aggressive layout is being introduced in more vehicles.

Performance of the brakes to decelerate the car and simultaneously maintain the steering capability is a very important factor. The FMVSS 105 and EEC Directive 71/320 describe many design and performance criteria. Each vehicle manufacturer also has its own additional test programs such as down-hill performance during a long descent from a high mountain and extensive winter testing.

Basic requirements are two-circuit brakes, a brake booster, a braking force regulator for the rear brakes, sufficient brake cooling, and multiple brakes out of high speed without a significant reduction of brake performance. Also, the perfect function of the brake system during the lifetime of the vehicle is a very important parameter for safe driving. As mentioned, the self-stabilizing effect at the front axle is one important element to keep the vehicle in line during braking on roads with different coefficients of friction from side to side of the vehicle. Figure 6.23 shows a cross section of the wheel contact area (midpoint), the spring strut, and the track rod.

The plan view shows the stability function. In a braking situation at the wheel with the higher brake force, a moment arises that corrects the tendency to steer in the direction of this wheel; thus, the vehicle track remains stable. For vehicles, it also is important that the rear wheels do not lock during brak-ing. This is achieved, when necessary, by a brake pressure reducer or with the ABS. State-of-the-art for the ABS is the three-channel system. Fig-ure 6.24 shows the schematic function. Via electromagnetic valves, the brake pressure is controlled at the single wheel to avoid the locking of one of the specific wheels.

The opinion often mentioned by the public is that an ABS reduces, in any case, the braking distance. However, this is not true. The main purpose of the ABS is the capability to steer and brake. The span of control of the ABS allows a variation of 8 to 35% slip between a single wheel and the road

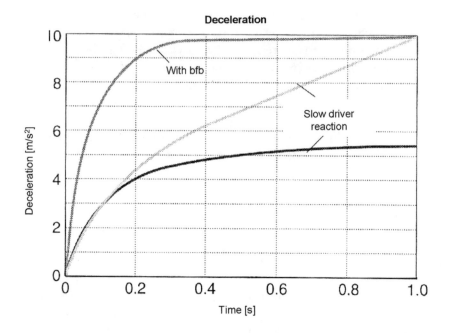

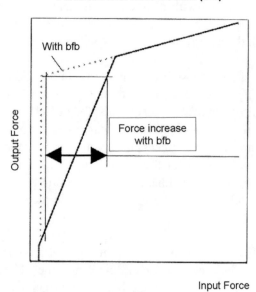

Figure 6.22 Brake assistant. (Source: Ref. 6-21.)

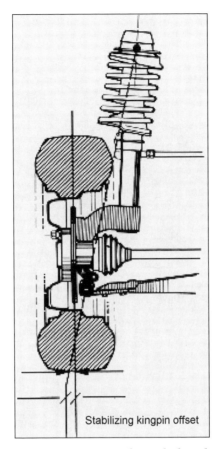

Stabilizing kingpin offset

*Figure 6.23 Cutaway view through the wheel–road
contact area. (Source: Ref. 1-1.)*

surface. With this data, the brake distance is optimally short, but there is
enough side force, which is necessary for steering.

6.3.3 Brake-by-Wire

The preceding examples have shown the importance of electronic brake management related to accident avoidance. Two systems are being investigated
to use brake-by-wire technology for brake systems. The first application in a

ABS *control cycle for high adhesion coefficients.*

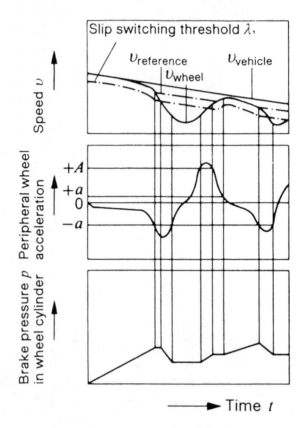

Figure 6.24 Function of an ABS control (one wheel).
(Source: Ref. 6-22.)

production car is the sensotronic brake control (SBC), an electro-hydraulic brake system (EHB) [6-23]. Figure 6.25 shows the general layout.

The systems contain the following components:

• Activation unit, including the brake pedal

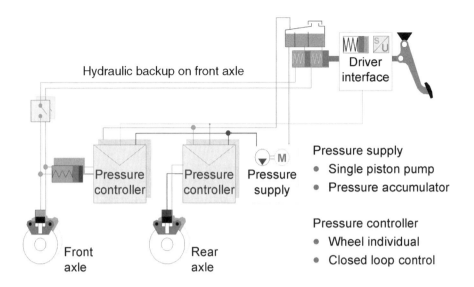

Pressure supply
• Single piston pump
• Pressure accumulator

Pressure controller
• Wheel individual
• Closed loop control

Figure 6.25 Sensotronic brake control (SBC). (Source: Ref. 6-23.)

• Hydraulic control unit with a specific electronic control unit, which could be installed in a location separate from the activation unit

• Yaw rate sensor, including a lateral acceleration sensor

• Rpm sensors at the wheels

• Steering wheel angle sensor

With this system, greater design freedom for the brake functions is possible. The result shows shorter stopping distances, reduced pedal effort, and a better feeling at the brake pedal (e.g., no vibration of the brake pedal as occurs with conventional ABS). Also, optimization of ABS, ASR, ESP (Electronic Stabilization Program), and ACC is much easier. Likewise, the system is more convenient through smart stop. It also will be one of the important features if automated driving on roads becomes available in the future. In an accident, it is advantageous that the tandem master cylinder used in the system has a shorter length in front of the brake pedal. Therefore, any intrusion into the passenger compartment could be minimized. Figure 6.26 shows

- Electronic brake pedal
  electronic control of the whole braking system

- Electrically driven power supply

- High-pressure accumulator

- Controllable ball valves

- Closed loop control of hydraulic pressure
  in the wheel brakes — wheel individual

- Hydraulic backup
  lowest degraded mode of the braking system

Figure 6.26 Optimized brake function by SBC. (Source: Ref. 6-23.)

the optimal brake pedal force versus the pedal movement. The deceleration versus pedal movement shows that braking must be a compromise between too "poisonous" and too "blunt" performance. If a system failure occurs, a fall-back situation is given through the fact that with the remaining hydraulic circuit, the two front wheels can be used to brake the vehicle.

With the SBC (EHB) system, several additional functions are possible. For example, in critical situations where emergency braking is necessary, the brake assistant increases the brake line pressure until the antilock braking controller reacts. Also, a pre-filling of the brake lines is possible in the case where the acceleration pedal is relieved with a jerking motion in anticipation of the need for braking. Even as the rain sensor senses the start of a rainfall, the braking system with unnoticeable brake impulses can remove a water or salt coating from the tires.

The second advanced system is the electronic mechanical brake system (EMB). Figure 6.27 shows the layout in principle.

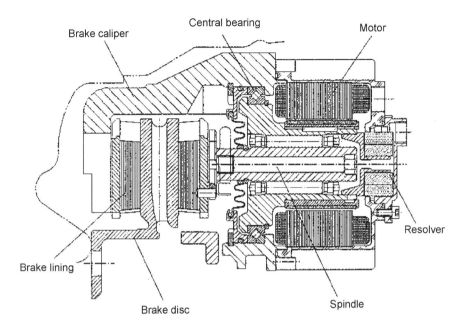

Figure 6.27 Cutaway view of an electronic mechanical brake system (EMB). (Source: Ref. 6-24.)

The basic difference with the EHB is that the hydraulic unit is replaced by an electric motor with a mechanical adjustment unit at each wheel to apply the brakes via an electrical signal. It is evident that the possible design freedom of this system is even greater. On the other hand, the control unit must be fail safe; that is, without electric power, the system should not be able to lock the brakes. In addition to this requirement, we need at least a second independent electric circuit, including the battery, to ensure that the brakes continue to perform if one electrical circuit does not work.

6.3.4 Vehicle Dynamics

The front-wheel-drive system has the highest market share compared to other vehicle layouts such as standard rear-wheel drive and four-wheel drive. The front-wheel-drive system offers the driver comfortable steering and a nicely tuned chassis, even in small vehicles. Also in this area, we find, in addition to hydraulic power steering, electrical devices to decrease the torque at the steering wheel. For a good straight-ahead drive, both the front and rear axles must be designed to fulfill the requirements for a good-natured, forgiving but precise and comfortable chassis. Figure 6.28 shows the front and rear axles of a modern compact passenger car [6-25].

At high speeds, it is necessary to minimize vertical lift at the rear axle. In the design of the vehicle, we must find a good compromise between low aerodynamic resistance and low rear-axle vertical lift. Figure 6.29 shows the aerodynamic coefficient, CD, versus the rear axle lift coefficient, CAH. If the rear-end lift is too high, especially during lane-changing maneuvers at high speed, this might become critical.

The correlation between spring and damping comfort and sufficient contact between tire and road also is important. This correlation was shown here previously in Figure 6.5. Although we have improved theoretical knowledge and better simulation tools, the knowledge of experienced test engineers is still used for chassis design and layout. Numerous tests, as shown in Table 6.5, are used to determine vehicle behavior under dynamic situations [6-26].

One special test in Europe was invented after much public discussion about the behavior of vehicles in extreme driving maneuvers. Although the ESP was already under development, the tests and the public discussion about a so-called "moose (elk) avoidance maneuver" definitely accelerated the installation of the ESP system in production cars. The ESP supplements the following systems already installed:

- Antilock brake system (ABS)
- Electronic brake booster (EBV)
- Electronic differential system (EDS)
- Antiskid control device (ASR)

The yaw moment controller uses the measured variables: wheel velocity, yaw velocity, steering wheel angle, lateral acceleration, and pre-pressure at the brake master cylinder. With the input of the steering wheel angle and the vehicle velocity, a setpoint value is determined and compared to the signal of the yaw velocity and lateral acceleration sensor. If any differences exist, the yaw moment control unit generates signals that control the brake pressure of

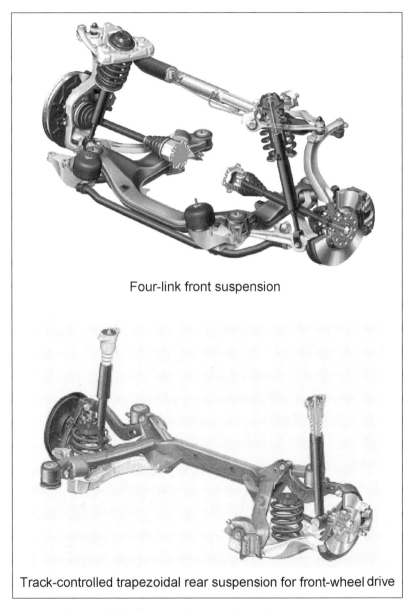

Four-link front suspension

Track-controlled trapezoidal rear suspension for front-wheel drive

Figure 6.28 Front and rear axles of a compact car.
(Source: Ref. 6-25.)

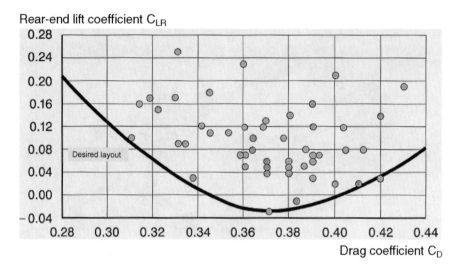

Rear-end lift coefficient C$_{LR}$

Drag coefficient C$_D$

*Figure 6.29   Rear-end lift coefficient, as a function of drag coefficient.
(Source: Ref. 6-4.)*

each individual wheel, as well as the engine and transmission ECU. The resulting forces at the single wheels allow stabilization of a vehicle that has a tendency to skid. The ESP is the first system that supports the driver to a large extent, compared with systems that can be judged to be activated by the driver. One design—known as ESP-Plus—uses three wheels to control the vehicle function. Figure 6.30 demonstrates the principle function.

Another system that offers a compromise between safety and vehicle dynamics is active chassis control, which also allows a high degree of driver comfort. For example, the active chassis could influence in a positive way the roll gradient of the vehicle [6-28].

Other factors are important in the design of accident avoidance means, such as the tire and rim. Figure 6.31 shows an advanced system that even includes a sidewall torsion sensor [6-29].

In addition to the specific performance requirements for tire and rim design, tire pressure is a key element with respect to fuel economy and vehicle safety. Because most customers do not carefully monitor tire pressure as frequently

TABLE 6.5
EXAMPLE OF SINGLE CRITERIA FOR CHASSIS CONTROL

Subjective Assessment of Driving Behavior

1. Drive-Away Behavior		4. Cornering Behavior	
1.1	Squat	4.1	Cornering behavior
1.2	Drive-away oscillation	4.2	Turn-in ability
1.3	Drive-away shaking	4.3	Lateral force increase
1.4	Steering input	4.4	Yaw velocity increase
1.4.1	Coefficient of friction: High	4.5	Transverse control ability
1.4.2	Coefficient of friction: Split	4.6	Roll behavior
1.5	Torque-steer	4.7	Diagonal dip
1.6	Steering jam	4.8	Support effect
1.7	Traction	4.9	Roll screening
1.7.1	Coefficient of friction: High	4.10	Lane-change performance
1.7.2	Coefficient of friction: Low	4.11	Steer-brake performance
1.7.3	Coefficient of friction: Split	4.12	Steer-acceleration performance
1.7.4	Coefficient of friction: Sudden change	4.13	Road impact
1.8	Control response ATC	4.14	Load alteration effect
1.9	Pedal rear travel ATC	**5. Straight-Running Stability**	
2. Braking Performance		5.1	Straight ahead
2.1	Braking Deceleration	5.2	Spring steering
2.1.1	Coefficient of Friction: High	5.3	Roll steering
2.1.2	Coefficient of friction: Low	5.4	Steer oscillation
2.1.3	Coefficient of friction: Split	5.5	Ridging
2.1.4	Coefficient of friction: Sudden change	5.6	Track rut sensitivity
2.2	Stability	5.7	Load alteration steering effect
2.3	Straight ahead stability	5.8	Side wind sensitivity
2.4	Cornering stability	5.9	Wind sensitivity
2.5	Steer stability	5.10	Trailer wobbling
2.6	Yaw stability	**6. Driving Comfort**	
2.6.1	Coefficient of friction: High	6.1.1	Ride comfort, low speed
2.6.2	Coefficient of Friction: Low	6.1.2	Ride comfort, high speed
2.6.3	Coefficient of friction: Split	6.2	Pitch behavior
2.7	Brake dive	6.3	Roll behavior
2.8	Pedal effort	6.4	Body damping
2.9	Pedal feeling	6.5	Rolling comfort
2.10	Pedal reaction ABS	6.6	Harshness
2.11	Pedal moving ABS	6.7.1	Rolling noise
2.12	Brake judder	6.7.2	Tire whining
2.13.1	Brake noise general	6.8	Edge sensitivity
2.13.2	Squeal	6.9	Roar
2.14	Tramp	6.10.1	Thumping
3. Steering Behavior		6.10.2	Damper
3.1	Pivoting	6.11	Bouncing
3.2	Responsiveness	6.12	Absorb capability (bumps)
3.3	Trench effect	6.13	Rebound
3.4	Center point	6.14	Buffer start
3.5	Steering effort	6.15	Return
3.5.1	Central position	6.16	Tire/freeway hop
3.5.2	Proportional range	6.17	Springing
3.5.3	Parking	6.18	Stutter (5–15 Hz)
3.6	Steering passing	6.19	Load alteration
3.7	Overshoot	6.20	Body vibration
3.8	Post-oscillation	6.21	Steering vibration
3.9	Post-steering	6.22	Steering shimmy
3.10	Target precision	6.23	Steering bouncing
3.11	Road contact	6.24	Steering return kick
3.12	Maneuverability	6.25	Steering rattle
3.13	Steering return	6.26.1	Seating comfort isolation
		6.26.2	Seat lateral support

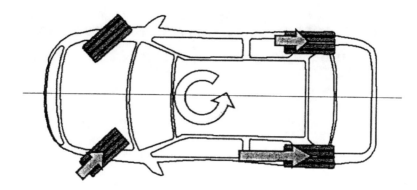

Figure 6.30 Principle of layout of an ESP system.
(Source: Ref. 6-27.)

Development of Modern Tires: The Tire as an Integrated Element
Advanced Tire Technology; Interaction Brake/Tire/Road

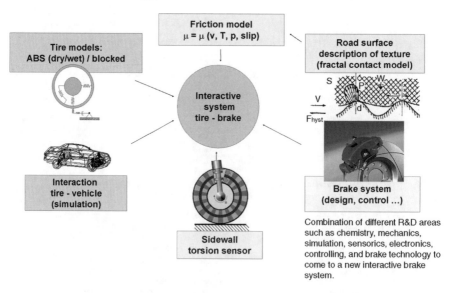

Figure 6.31 Sidewall torsion sensor. (Source: Ref. 6-29.)

as they should, the tire pressure will be reported to the driver by sensors installed at each wheel. Different technical solutions are under development. For example, if the tire pressure becomes too low, radio signals transmit the signals to the instruments in the dashboard. One example is shown by the Beru RDKS, which also offers an aftermarket solution. This system consists of wheel electronics, valves, sensors, antennas, and an ECU [6-30].

It is evident that all electronic and drive-by-wire systems must be more precise in their development with respect to software—and hardware—security, as described, for example, in Ref. 6-31. In the field of accident avoidance, many innovations are already in production vehicles available today, as shown in Figure 6.32 [6-32].

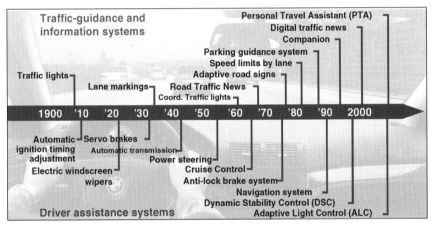

Figure 6.32 Driver assistance and traffic guidance systems: a chronology. (Source: Ref. 6-32.)

The future will hold even more systems such as active lights, active steering systems, active gas pedal, and extended information systems such as floating car data and car-to-car communication. As mentioned, these systems will be successful only if the man-machine interface is designed in such a way that the driver is integrated into the appropriate loop while retaining control of the vehicle. This was demonstrated in a positive way, in a demonstration vehicle by Delphi [6-33], although actual production is years away. This Delphi system included the following features:

The integrated safety system (ISS) defines increased safety in a broader approach and combines collision avoidance and mitigation of injury systems with electronics. The key to future vehicle systems in this regard is the successful integration of the following technologies:

- Positive solutions for the man-machine interface

- Functional multimedia systems, such as telematics

- Driver workload management (attention)

- 360° sensors for collision warning/intervention systems (e.g., lane change, blind spot, and vehicle control), adaptive cruise control (ACC), and/or lateral guidance

- Control of the chassis component (comfort and accident avoidance)

- Advanced safety interiors (smart restraints)

- X-by-wire technology

6.4   Information Systems

The information system as an amendment to driver assistance technology contributes to possible accident reduction to a large extent. This means not only visible roads but uniform, understandable, and clear traffic signs and the integration of the driver and vehicle into an information and communication system. Figure 6.33 shows the introduction of this type of support system for the future [6-13].

With navigation systems supported by traffic data, the driver can identify traffic jams much earlier and thus avoid them. In the future, it also will be possible to inform the driver about direct events, such as fog, icy roads, accidents, and stopped traffic in a critical curve in the road. Other features include "hands-free" mobile phones, emergency calls, and services such as e-mail or fax, naturally not while at the driver's seat or during driving. On a national and international basis, much research is being done to improve these systems and to define standards that allow the function, even if we change the border

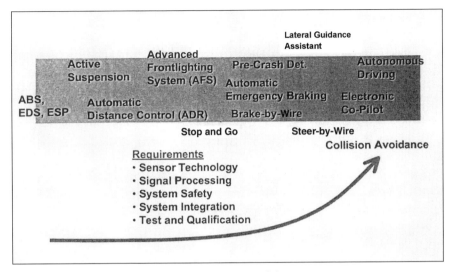

Figure 6.33 Evolution of a road map of driver assistance systems.
(Source: Ref. 6-13.)

from one country to another. These programs are performed on an international basis.

In Europe, these programs include the following:

- Information Society Technologies (Europe) (IST) (user friendly in the information society)

- E-commerce

- ISIS (Information Society Initiative in Standardization)

- European Space Agency (ESA), new-generation spaceborne Global Positioning System (GPS), navigation, and mobile services

- ERTICO, reliable travel and information; reliable and user-friendly public transport; traffic management ensures efficient use of transport; faster emergency services save lives; easy cash-less payment; high-quality and cost-effective freight and fleet services; driver assistance systems for safer vehicles and safer roads; and beyond—the automated highway

Programs in the United States and Japan include the following:

• Society of Automotive Engineers (SAE) (active), accident prevention

• U.S. Department of Transportation, safety

• IVHS, electronic toll industry, commercial vehicle operations, traffic management, border crossing, parking (access control), transit

The international programs are supported by national research [6-34]. With more advanced communication systems, better integration into a total traffic management system is possible. This will allow the following:

• Planning of the driving route, also from home

• Commercial fleet management, optimal use of roads and traffic means (e.g., cars, rail, buses)

The responsible traffic planner also has an opportunity to shape the traffic flow. If we look further into the future, other technical features will be possible, such as:

• Pre-crash determination
• Automatic emergency braking
• Steer-by-wire
• Collision avoidance
• Electronic co-pilot
• Autonomous driving

It is inefficient to optimize only the vehicle. We also must make the road more intelligent (see Figure 6.34) [6-35, 6-36]. Some items for this area are the following:

• Increased visibility
• Warning in advance of hazardous roads
• Vehicle counting and classification
• Identification of traffic congestion
• Obstacle and accident warning

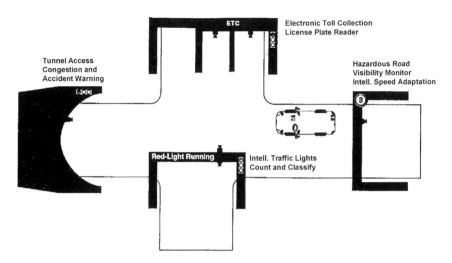

Figure 6.34 Mobility with vision. (Source: Ref. 6-35.)

- Intelligent traffic light and speed adaptation
- Limited tunnel access

The date of the introduction of autonomous driving cannot be determined today. It might be appropriate to start this technique not in normal public traffic but in special applications. Nevertheless, the other systems described will be introduced into the field via a step-by-step approach. The pre-conditions for this are driver acceptance, optimization of the man-machine interface, and achievement of an overall benefit.

6.5 References

6-1. Bundesministerium für Verkehr, Bau und Wohnungswesen (Ed.), *Verkehr in Zahlen 2001/2002*, Vol. 30, Deutscher Verkehrsverlag, Hamburg, ISBN 3-87154-270-9.

6-2. Busch, S., Schwarz, T., and Zobel, R. "Determination of Risk Factors of Accident Causation," Institution of Mechanical Engineers Conference on Vehicle Safety, May 28–29, 2002, London, Volkswagen AG.

6-3. Behr GmbH & Co., Stuttgart. Klimatisierung [Air-Conditioning], VDA-Report, September 2001, Frankfurt am Main, Germany.

6-4. Braess, H.-H. and Seiffert, U. *Handbuch Kraftfahrzeugtechnik*, Vieweg-Verlag, Wiesbaden, Germany, 2001.

6-5. Richter, B. *Schwerpunkte der Fahrdynamik*, Verlag TÜV Rheinland GmbH, Köln, 1990, ISBN 3-88585-772-3.

6-6. Becker, Axel, et al. "Maßnahmen zur Verkürzung des Anhalteweges in Notbremssituationen—das '30 m Auto,'" Proceedings, Verband der Automobilindustrie (Association of the [German] Automobile Industry) Technical Congress, March 26–27, 2001, Bad Homburg v. d. Höhe, Germany.

6-7. Peters, M. "Bedienkonzept im Fahrzeug—Das intuitive und adaptive HMI," Forschungs- und Entwicklungszentrum—MMI—BMW AG, München, Germany.

6-8. Braess, H.-H. and Seiffert, U. *Handbuch Kraftfahrzeugtechnik*, Vieweg-Verlag, Wiesbaden, Germany, 2001.

6-9. National Highway Traffic Safety Administration (NHTSA) and Federal Motor Vehicle Safety Standard (FMVSS), FMVSS 103, Windshield Defrosting and Defogging Systems, Washington, DC, United States.

6-10. National Highway Traffic Safety Administration (NHTSA) and Federal Motor Vehicle Safety Standard (FMVSS), FMVSS 104, Windshield Wiping and Washing Systems, Washington, DC, United States.

6-11. Society of Automotive Engineers, SAE J902, "Passenger Car Windshield Defrosting Systems," SAE International, Warrendale, PA, United States.

6-12. Robert Bosch GmbH, Stuttgart. Product information, undated.

6-13. Specks, W. "Elektronikkonzepte für zukünftige Fahrerassistenz-systeme," Verband der Automobilindustrie (Association of the [German] Automobile Industry) Tehcnical Congress, March 26–27, 2001, Bad Homburg v. d. Höhe, Germany.

6-14. Knoll, P. "Surround Sensing—Collision Warning Systems—Vehicle Guidance," 01A2002, Bosch Group, Möglingen, Germany.

6-15. Hella KG Hueck & Co. Product information, 2002, Lippstadt, Germany.

6-16. Eichhorn. U. "Assistenzsysteme für das Auto der Zukunft," Fortschritt-Berichte VDI Series 12, No. 485, Düsseldorf, Germany, 2002.

6-17. Robert Bosch GmbH (Ed.). *Kraftfahrtechnisches Taschenbuch*, 22nd ed., VDI-Verlag, Düsseldorf, 1995, ISBN 3-18-419122-2.

6-18. Pape, E. "VW in 4 Motion," Proceedings of the Four-Wheel Drive Congress, February 14–15, 2002, Graz, Austria.

6-19. Knoll, P. Surround Sensing and Sensor Data Fusion Technical Congress, March 2002, Stuttgart, Germany.

6-20. Author's unpublished data. See also Zobel, R. "Analyse des realen Unfallgeschehen Methoden und Prinzipien der VW-Unfallforschung," Proceedings of the conference "Kollisionsschutz im Straßenverkehr," November 6–7, 1995, Haus der Technik, Essen, Germany.

6-21. Braess, H.-H. and Seiffert, U. *Handbuch Kraftfahrzeugtechnik*, Vieweg-Verlag, Wiesbaden, Germany, 2001.

6-22. Robert Bosch GmbH (Ed.). *Automotive Handbook*, ISBN 0-89 283-518-6.

6-23. Booz, Othmar, et al. "Electro-Hydraulic Brake with the Focus on the Electric Power Supply," Verband der Automobilindustrie (Association of the [German] Automobile Industry) Technical Congress, March 2002, Stuttgart, Germany.

6-24. Belschner, R. et al. *Brake by Wire Using a TTP/C Communication Network*, VDI-Berichte 1415, Electronic Systems for Vehicles, VDI-Verlag, Düsseldorf, 1998, ISBN 0083-5560.

6-25. Heißing, B., et al. "The New Audi A4," Special Edition, *ATZ/MTZ*, 2000, Vieweg, Wiesbaden.

6-26. Heißing, B. and Brandl, H.J. *Subjektive Beurteilung des Fahrverhaltens*, Vogel-Verlag, Würzburg, Germany, 2002, ISBN 3-8023-1906-6.

6-27. "Besser im Griff," *Automobil Industrie*, May 2002, Germany.

6-28. Berkner, S. et al. "Beeinflussung der Querdynamik vom Pkw durch aktive Fahrwerke, HDT." Conference on Driveability: Fahrkomfort, Fahrspaß, und Fahrsicherheit, Essen, June 26–27, 2001.

6-29. Huinink, H. et al. "Dynamische Interaktion Bremse—Reifen—Straße," 18 µ Symposium, October 23–24, 1998, Bad Neuenahr, Germany; in: Fortschritt-Berichte VDI, Series 12, No. 373, VDI-Verlag, Düsseldorf, Germany, 1998.

6-30. Beru Aktuell, "The Beru RDKS," January 2002, Ludwigsburg, Germany.

6-31. Binfet-Krull, M. et al. "Definition of Safety/Reliability Requirements for Components of Electronic Vehicle Systems Like Steer by Wire," 4th European Conference and Exhibition, Coventry, June 27–28, 2001.

6-32. Frank, D., et al. "Wo liegen die Grenzen der Fahrerassistenz?" Proceedings, VDA Technical Congress, March 2001, Bad Homburg, Germany, Frankfurt 2001.

6-33. Delphi Automotive Systems. Product information material during the IAA 2001, Frankfurt, Germany.

6-34. Kasties, G. "Verkehrstelematik: Staus müssen nicht sein," in *Die heimlichen Siegerbranchen*, FAZ Buch Frankfurt, 2002, ISBN 3-89843-008-1.

6-35. Hoefflinger, B. *Mobility with Vision*, Institute for Microelectronics, Stuttgart, Germany, 2001.

6-36. Insurance Institute for Highway Safety annual status report, Vol. 37, No. 5, May 4, 2002, Arlington, VA, United States.

7.

Biomechanics and Occupant Simulation

7.1 Definition

For mechanical engineers, it might be surprising that the discipline of biomechanics developed and that it has become so important to develop vehicles by understanding injury mechanisms. Biomechanics can be called the science that applies the principles of mechanics to biological systems [7-1]. It is not a new research discipline. Galileo (1564–1642) and Harvey (1578–1658) worked in this field during their time. Biomechanics includes in-depth studies on the behavior of humans under internal and external forces, as well as applied engineering work. Therefore, it is not surprising that many different disciplines are included in biomechanics: engineering, epidemiology, traumatology, anatomy, biology, and physiology.

In the last 50 years in the United States, Europe, and Japan, research activities in this area have increased. The necessary link between medical experts and engineering was created because many organizations in various countries openly exchange their research results at conferences such as the Experimental Safety Vehicles (ESV), Stapp Car Crash, International Research Council on the Biomechanics of Impact (IRCOBI), Society of Automotive Engineers (SAE), and others, as well as by personal contacts. Furthermore, large vehicle manufacturers and insurance companies are supporting work in this field.

An important contribution to knowledge about the resistance of humans against a high-rate impact load was made by an American, Colonel John Stapp, who personally served as the first human to be decelerated from 632 mph ($\approx$1000 km/h) in 1.4 sec to zero. If we take a rectangular deceleration pulse, this corresponds to approximately 20g. The annual Stapp Car Crash Conference therefore is named after Colonel John Stapp. At the Eighth Stapp Conference in 1964 [7-2], the following appreciation to Colonel Stapp was made:

> The Stapp Car Crash Conferences are named in honor of Colonel John Stapp, USAF (MC), who pioneered (and is still pioneering) in establishing human impact tolerance levels. His historic rocket sled rides at Holloman Air Force Base, New Mexico, in 1954, in which he voluntarily subjected himself to up to 40g accelerations while stopping from a speed of 632 miles per hour in 1.4 seconds, still represent the best basis for quantifying human tolerance to acceleration. In addition to his own dangerous volunteer work, he has directed countless other safety research programs involving human volunteers, animals, and cadavers. The equipment and techniques developed under his guidance have become standard in this research area and have contributed much to the advancement of safety. The naming of these conferences after Colonel Stapp is a fitting tribute to a man who has dedicated his life—even to the point of risking it—to research aimed at increasing man's chances of survival in adverse crash environments.
>
> The conferences were initiated at the University of Minnesota (Colonel Stapp's alma mater) under the able direction of Professor James J. Ryan, another outstanding researcher in crash safety. For four years, the conferences were held at either the University of Minnesota or an appropriate U.S. Air Force base. Currently, the conference rotates annually among four sponsors: The University of Minnesota (1961), the United States Air Force (1962), the University of California at Los Angeles (1963), and Wayne State University (1964). The 1965 meeting will again be held at the University of Minnesota on October 20, 21, and 22. The proceedings of the conference are published in bound form and will, it is hoped, become a valuable reference source." [7-2]

Colonel Stapp is only one of the pioneers. Around the world, many others are active in this type of research. In connection with vehicle safety of traffic participants, the biomechanical results also are an instrument for determining the biomechanic limits of humans. The results of biomechanic research lead to the definition of load limitations. From that, protection criteria are taken, which should serve as limits that should not be exceeded. For this reason, we first need to know what is happening in real-world accidents and which injury mechanisms are important.

7.2 Injury Tolerance Limits

Injury tolerance limits describe items such as fractures, injuries of organs, and other injuries. A classification is done via the Abbreviated Injury Scale (AIS) or Overall Abbreviated Injury Scale (OAIS). With the AIS or OAIS, the single or total injury is described. The data span a range from 0 to 6. Table 7.1. shows the severity rating versus AIS. To more clearly illustrate the relationship between injury description and AIS, Table 7.2 provides an interesting description.

TABLE 7.1
THE ABBREVIATED INJURY SCORE (AIS) [7-3]

AIS	Severity Code
0	No injury
1	Minor
2	Moderate
3	Serious
4	Severe
5	Critical
6	Maximum injury (virtually unsurvivable)
9	Unknown

The limits from the injury level depend on age, sex, anthropometrics, mass, mass distribution, and specific conditions. This is why it is relatively complicated to cover all traffic participants in accident simulation tests—from

TABLE 7.2
AIS EXAMPLES BY BODY REGION [7-4]

AIS	Head	Thorax	Abdomen and Pelvic Contents	Spine	Extremities and Bony Pelvis
1	Headache or dizziness	Single rib fracture	Abdominal wall; superficial laceration	Acute strain (no fracture or dislocation)	Toe fracture
2	Unconscious <1 hr; linear fracture	2–3 rib fracture; sternum fracture	Spleen, kidney, or liver; laceration or contusion	Minor fracture without any cord involvement	Tibia, pelvis, or patella; simple fracture
3	Unconscious 1–6 hr; depressed fracture	≥4 rib fracture; 2–3 rib fracture with hemoth. or pneumoth.	Spleen or kidney; major laceration	Ruptured disc with nerve root damage	Knee dislocation; femur fracture
4	Unconscious 6–24 hr; open fracture	≥ 4 rib fracture with hemoth. or pneumoth.; fail chest	Liver; major laceration	Incomplete cord syndrome	Amputation or crush above knee; pelvis crush (closed)
5	Unconscious >24 hr; large hematoma (100 cc)	Aorta laceration (partial transection)	Kidney, liver, or colon rupture	Quadriplegia	Pelvis crush (open)

vehicle occupants via cyclists to pedestrians. Therefore, it can be observed that over recent years, even more test devices or three-dimensional dummies or computer models have been used.

7.3 External Injuries

A rating of cuts in faces was done by Professor L. Patrick of Wayne State University. Because of the laminated windshield, seat belts, and airbags, this type of injury is significantly reduced. The breaking of the skull due to impact onto rigid parts was described in Ref. 7-5 as follows: The deceleration on the head multiplied with the head mass results in a force that could break part of the head. Figure 7.1 shows mechanical values for parts of the human body and the injury level.

Body Part	Mechanical Variables	Load Values
Total Body	$a_{x\,max}$ $\bar{a}_x$	40...80g 40...45g, 160...220 ms
Brain	$a_{x\,max}$, $a_{y\,max}$	100...300g WSU-curve with 60g, T>45 ms 1800...7500 rad/s²
Skull Fracture	$a_{x\,max}$, $a_{y\,max}$	80...300g depending on the size of the impact area
Forehead	$a_{x\,max}$ F_x	120...200g 4000-6000 N
Cervical Spine	$a_{x\,max\,thorax}$ $a_{y\,max\,thorax}$ F_x $\alpha_{max\,forward}$ $\alpha_{max\,rearward}$	30...40g 15...18g 1200...2600 N shear force 80°...100° 80°...90°
Thorax	$a_{x\,max}$ F_x S_x	40...60g, t>3 ms 60g, t<3 ms 4000...8000 N 5...6 cm
Pelvis-Femur	F_x $A_{y\,amx}$	6400...12500 N force application in the femur 50...80g (pelvic)
Tibia	F_x E_x M_x	2500...5000 N 150...210 Nm 120...170 Nm

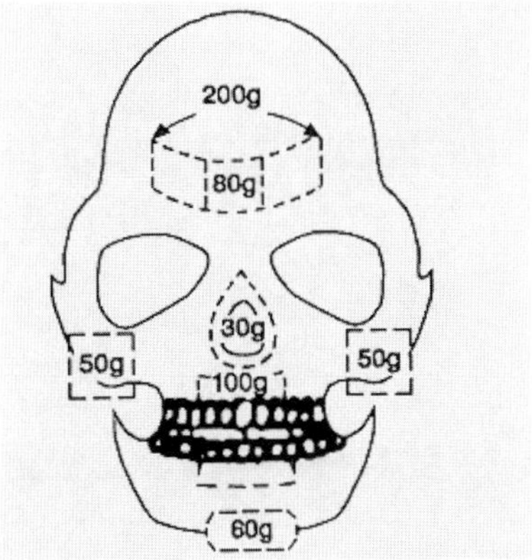

Figure 7.1 Biomechanical limits on humans. (Source: Ref. 7-5.)

For example, the limits for the forehead are 80g, the nose 30g, and the chin 40g. As mentioned, this data could not be used directly if dummies were taken for the analysis because dummies are not a precise duplication of humans.

7.4 Internal Injuries

The injury mechanism for internal parts of the human body is much more complicated to determine. The biggest problem is, without a doubt, the load on the brain and the cervical vertebra. For the head, the tolerance level is the limitation of the g-level in the anterior–posterior direction, with a value of 80g over a time period of more than 3 ms that should not be exceeded.

7.4.1 Concussion

A basic work on this subject was conducted by Wayne State University under the supervision of Professor L. Patrick [7-6]. In this experiment, embalmed cadavers were used and impacted a rigid or padded flat plate. Figure 7.2 shows the results of these tests.

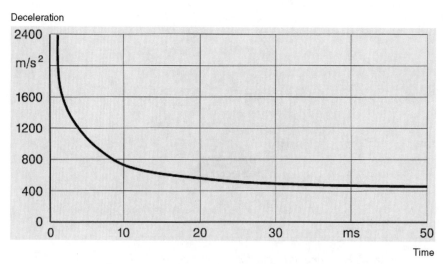

Figure 7.2 Patrick curve (scale for the evaluation of loadings on the human brain).

It can be stated clearly that this is the basis for many rule-making decisions in setting up the specified limits. If we analyze the curve shown in Figure 7.2, it can be easily understood that there is a g-level time relationship. For this reason, the severity index (SI) was developed, which was calculated by the following formula:

$$SI = \int_{O}^{T} a^{2.5}(t)\,dt$$

From a biomechanical viewpoint, a maximum level of 1000 should not be exceeded for a unidirectional acceleration measurement. Later, a new criteria was established, which is known in the United States as head injury criteria (HIC) and in Europe as head protection criteria (HPC).

$$HIC = \left(\frac{1}{t_2 - t_1} \int_{t_1}^{t_2} a_{res} \cdot dt \right)^{2.5} \cdot (t_2 - t_1) < 1000$$

where t is in seconds, a is the resultant acceleration measured in the head, and t_1 and t_2 are arbitrary time points. The HIC is calculated by an iterative mathematical model in such a way that the measured acceleration time function is achieving a maximum. To demonstrate the influence of the shape of acceleration time function, three different characteristic curves are shown in Figure 7.3.

Although the maximum value of all three curves has the same level of 100g, the relevant HIC calculated has values of 246, 419, and 1000. The HICs mentioned are always the maximum values for the specific deceleration time function time t_1 and t_2, as determined by an iterative calculation for all possible t_1 and t_2, insofar that a maximum level is achieved. In addition to the fact that the HIC has only a limited information value (e.g., the HIC does not consider the rotational influence and is not based on resultant acceleration), we agree that this formula and the limit of 1000 (–) is used on a worldwide basis to judge the performance of vehicles in accident simulation tests.

Resultant deceleration

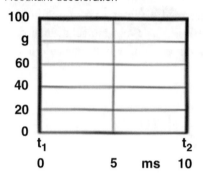

HIC	= 1000
t_1	= 0 ms
t_2	= 10 ms
ā	= 100g

Resultant deceleration

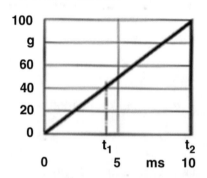

HIC	= 246.4
t_1	= 4.3 ms
t_2	= 10 ms
ā	= 71.5g

Resultant deceleration

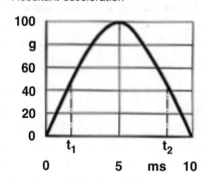

HIC	= 418.661
t_1	= 1.6 ms
t_2	= 8.3 ms
ā	= 82.8543g

Figure 7.3 Head injury criteria (HIC), as a function of three different deceleration time curves.

The rotational acceleration in the anterior–posterior direction was already investigated by Fiala [7-7] in the late 1960s. His investigation showed that with a brain mass of 1300 g, a value of 7500 rad/s^2 should not be exceeded.

7.4.2 Spinal Injuries

Another important field is the prevention of spinal injuries. Early in 1971, Patrick and Mertz determined the torque moments for flexion and extension via volunteers and human cadaver testing [7-8]. Figure 7.4 shows some

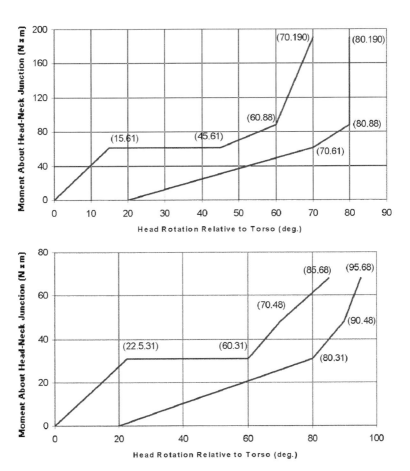

Figure 7.4 Head–neck response envelopes in flexion and extension for the loading phase. (Source: Ref. 7-8.)

head–neck responses. Table 7.3 shows some more recent results about maximum torque and forces related to the measurement with human volunteers [7-9].

TABLE 7.3
MAXIMUM STATIC FORCES AND BENDING TORQUES
DEVELOPED AT THE OCCIPITAL CONDYLES
BY HUMAN VOLUNTEERS [7-9]

	Bending Torque [Nm]
Forward Flexion	50.2
Extension	20.3
Lateral Flexion	47.5
	Force [N]
Anterior-Posterior (Shear)	845
Posterior-Anterior (Shear)	845
Lateral Shear	400
Axial Tension	1134
Axial Compression	1112

7.4.3 Chest Injuries

The chest is another area of critical injuries. It is evident that deformation force, deformation speed, and the deformation itself have an influence on chest injuries. Figure 7.5 shows force deflection measurements on unembalmed cadavers [7-10].

Where the force level has an average value of approximately 3 kN, deformation is in the range of 60 to 100 mm (2.4 to 4 in.). In lateral impacts, we found the force deflection curve as shown in Figure 7.6 [7-11].

It also was determined that deflection alone was not enough to determine possible injuries. This was why Viano established the viscous criteria (VC) [7-12], with the results as illustrated in Figure 7.7.

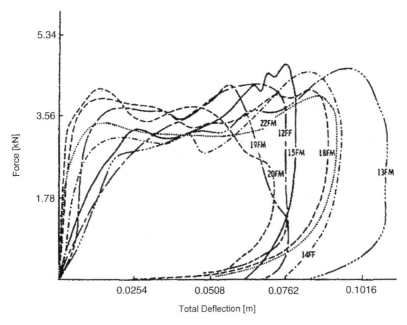

Figure 7.5 Dynamic force deflection in frontal impacts.
(Source: Ref. 7-10.)

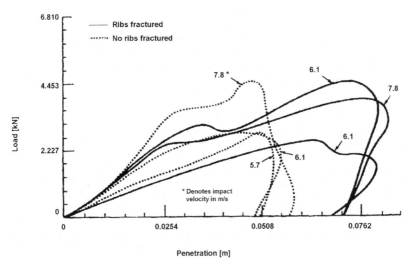

Figure 7.6 Dynamic force deflection in lateral impacts with padded
armrest-simulating impactor. (Source: Ref. 7-11.)

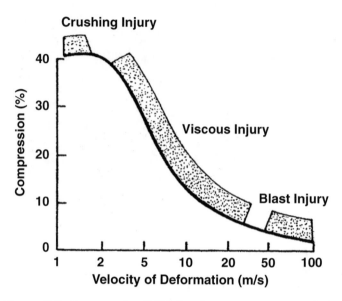

Figure 7.7 Ranges of validity for the viscous criteria (VC).
(Source: Ref. 7-12.)

7.5   Criteria in the Rule-Making Process

The basic requirements are defined in FMVSS 208 [7-13], the EEC directive
for frontal impacts [7-14], and the EEC directive for lateral impacts [7-15].

7.5.1 Head Protection

Head protection criteria, HIC or HPC, should not exceed 1000 or 700 and are
calculated as follows:

a. 1.   For any two points in time, t_1 and t_2, during the event which are
     separated by not more than a 36-ms time interval, and where t_1 is
     less than t_2, the head injury criterion (HIC_{36}) shall be determined
     using the resultant head acceleration at the center of gravity of the
     dummy head, a_r, expressed as a multiple of g (the acceleration of
     gravity), and shall be calculated using the expression

$$\left[\frac{1}{\left(t_2 - t_1\right)} \int_{t_1}^{t_2} a_r dt\right]^{2.5} \left(t_2 - t_1\right)$$

2. The maximum calculated HIC_{36} value should not exceed 1000.

b. 1. For any two points in time, t_1 and t_2, during the event which are separated by not more than a 36-ms time interval, and where t_1 is less than t_2, the head injury criterion (HIC_{15}) shall be determined using the resultant head acceleration at the center of gravity of the dummy head, a_r, expressed as a multiple of g (the acceleration of gravity), and shall be calculated using the expression

$$\left[\frac{1}{\left(t_2 - t_1\right)} \int_{t_1}^{t_2} a_r dt\right]^{2.5} \left(t_2 - t_1\right)$$

2. The maximum calculated HIC_{15} value should not exceed 700.

7.5.2 *Chest Protection*

Chest Deflection.

- VC = viscous criteria <1 m/sec

  a. Compressive deflection of the sternum relative to the spine shall not exceed 76 mm (3 in.).

- Lateral rip deflection <42 mm (1.65 in.)

  a. Compressive deflection of the sternum relative to the spine shall not exceed 63 mm (2.5 in.).

  b. Compressive deflection of sternum relative to the spine shall not exceed 50 mm (2 in.) (European requirement).

7.5.3 Neck Injury

When measuring neck injury, each of the following criteria shall be met:

1. The shear force (F_x), axial force (F_z), and bending moment (M_y) shall be measured by the dummy upper neck load cell for the duration of the crash event as specified. Shear force, axial force, and bending moment shall be filtered for N_{ij} purposes at SAE J211/1 rev. March 95 Channel Frequency Class 600.

2. During the event, the axial force (F_z) can be either in tension or compression, while the occipital condyle bending moment (M_{ocy}) can be either in flexion or extension. This results in four possible loading conditions for N_{ij}: tension–extension (N_{te}), tension–flexion (N_{tf}), compression–extension (N_{ce}), or compression–flexion (N_{cf}).

3. When calculating N_{ij} using the equation in (4) below, the critical values, F_{zc} and M_{yc}, are:

    $F_{zc} = 6806$ N (1530 lbf) when F_z is in tension

    $F_{zc} = 6160$ N (1385 lbf) when F_z is in compression

    $M_{yc} = 310$ Nm (229 lbf-ft) when a flexion moment exists at the occipital condyle

    $M_{yc} = 135$ Nm (100 lbf-ft) when an extension moment exists at the occipital condyle

4. At each point in time, only one of the four loading conditions occurs. The N_{ij} value corresponding to that loading condition is computed, and the three remaining loading modes shall be considered a value of zero. The expression for calculating each N_{ij} loading condition is given by

$$N_{ij} = \left(F_z/F_{zc}\right) + \left(M_{ocy}/M_{yc}\right)$$

5. None of the four N_{ij} values shall exceed 1.0 at any time during the event.

6. Peak tension force (F_z), measured at the upper neck load cell, shall not exceed 4170 N (937 lbf) at any time.

7. Peak compression force (F_z), measured at the upper neck load cell, shall not exceed 4000 N (899 lbf) at any time.

   Unless otherwise indicated, instrumentation for data acquisition, data channel frequency class, and moment calculations are the same as given for 49 CFR Part 572, Subpart E Hybrid III test dummy.

8. Neck rearward torsion moment should not exceed 57 Nm (requirement for Europe).

9. Force in longitudinal direction (requirement for Europe), $F_{ver} < (1.1–3.1$ kN) as f(t) (see Figure 7.8).

10. Neck shear force (requirement for Europe) $F_{shear} < (1.1–3.1)$ kN as f(t) (see Figure 7.9).

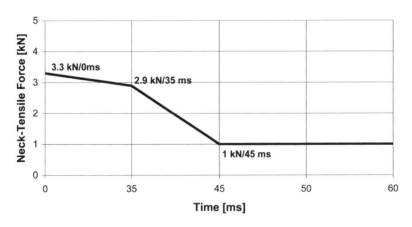

Figure 7.8 Neck tension load as a function of time.

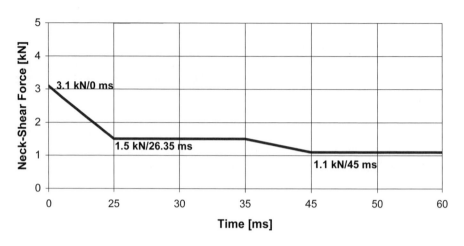

Figure 7.9 Neck shear force as a function of time.

7.5.4 Performance Criteria for the Rule-Making Process

Of many possible values, the following limits were established in the rule-making process.

7.5.4.1 Chest

Resultant chest acceleration <60g (>msec)

TTI = Thoracic Trauma Index   <85g (four-doors)
                                      <90g (two-doors)

TTI = 0.5 × (RIBY + T12Y)

RIBY = Maximum absolute value of lateral acceleration in g's of the fourth and eighth rib in the struck side

T12Y = Maximum absolute value of lateral acceleration in g's of the twelfth thoracic vertebra after filtering of the acceleration signal

Force < 8000 kN

7.5.4.2 Pelvic

Resultant acceleration <130g

Force abdomen <2.5 kN

Force symphysis <10 kN

7.5.4.3 Leg and Knee

Upper Leg:

Force limit in frontal impacts <10,000 N (requirement FMVSS 208)

Shear load in the knee joint <5000 N

Knee dislocation <15 mm (requirement for Europe)

Force in the femur in the longitudinal direction (requirement for Europe), $F_{long} < (7.6–9 \text{ kN})$ as f(t) (see Figure 7.10)

Lower Leg:

Force in the tibia (compression force criteria) (requirement for Europe) $F_{long} < 8 \text{ kN}$

Tibia index Ti measured at the top and bottom of each tibia must not exceed 1.3 at either location (requirement for Europe)

The tibia index is calculated on the basis of the bending moments (M_x and M_y) by the following expression:

$$TI = \left| M_R / (M_c)_R \right| + \left| F_z / (F_c)_z \right|$$

where

M_x = Bending moment about the x-axis

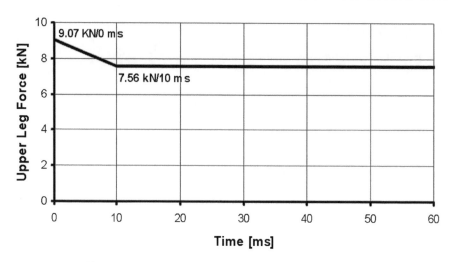

Figure 7.10 Upper leg force as a function of time.

M_y = Bending moment about the y-axis

$(M_c)_R$ = Critical bending moment and shall be taken to be 225 Nm

F_z = Compressive axial force in the z direction

$(F_c)_z$ = Critical compressive force in the z direction and shall be taken to be 35.9 kN

$$M_R = \sqrt{(M_x)^2 + (M_y)^2}$$

The tibia index is calculated for the top and bottom of each tibia; however, F_z may be measured at either location. The value obtained is used for the top and bottom TI calculations. Moments M_x and M_y are measured separately at both locations.

Due to the requirements already defined and because of improved restraint systems, especially the combination of seat belts and airbags, the degree of protection of the human body has increased. This means that the knee and the lower leg area have become a relatively high priority. For the feet of the front occupant, it is important to prevent too high a bending–flexion around the y-axis.

7.5.4.4 Pedestrian Protection

A special case are the limits proposed for the protection of pedestrians [7-16]. Figure 7.11 shows the different requirements:

a. Limitation of the headform acceleration for two different head forms, child and adult simulation

b. A simulation of a leg impacting the bumper of the vehicle front, where the leg angle should not exceed 15°, the shear distance of the knee should be below 6 mm (0.23 in.), and the acceleration of the tibia is less than 150g.

c. The simulation of the upper leg against the bonnet of the front hood with the requirement that the shear force should be lower than 4 kN and bending moment of the impactor less than 200 Nm.

7.6 Test Devices

Simulation of parts of the human body and the complete human is necessary to test vehicle components and the vehicle. Although professional investigations with human volunteers have been used, these can be performed only in situations where no injuries occur. From the experience of one of the authors at an age of 32 years, it can be stated that the level of the crash test should not be high enough to cause injuries. In a lateral car-to-car crash with an impact speed of 34 km/h (21 mph), this corresponds to a change in velocity of 17 km/h (10.5 mph), restrained by a standard three-point seat belt. The following observations could be made. The pulse frequency jumped from 105 to 175 during the impact phase, a small concussion could be observed, and muscle pain occurred over the whole body. This clearly was below the injury limit, but a slightly higher change of velocity already might have created minor injuries. For the development of vehicles and vehicle components, engineers must have reliable and reproducible results. Likewise, the test or

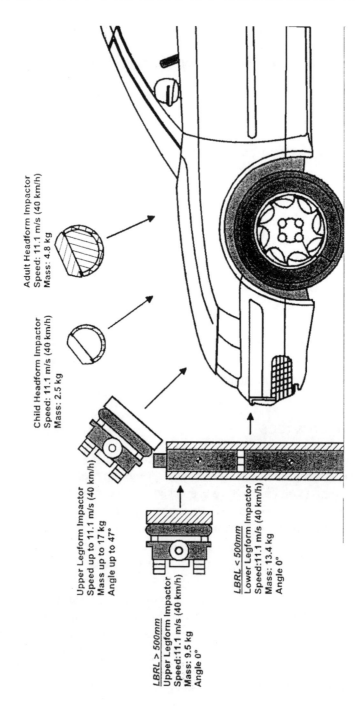

Figure 7.11 Test requirements for pedestrian protection. (Source: Ref. 7-17.)

calculations must be performed in areas where injuries can occur. This automatically prohibits the use of human volunteers as test specimens. For this reason, many test devices are available and are used for the development and evaluation of production cars.

7.6.1 Body Part Test Devices

For vehicles, component test devices are used in the following areas:

- The simplest equipment is a measurement device to control the radius of the outer parts of the vehicle. It must be greater than 3.2 mm (0.125 in.).

- Another measurement is done by a head form, which defines the impact zone at the dashboard and the vehicle interior.

7.6.1.1 Head Impact

Figure 7.12 shows a head impact test device that is used to measure the g-level during an impact of the reduced pendulum mass of 6.8 kg (15 lb).

The head impact form can be equipped with a three-dimensional accelerometer. It also can be used by attaching it with glue to foil, which shows the surface pressure if contact with a vehicle interior part occurs during the crash. The head form also could be used in a free-flying mode to check the deceleration level of the head form during contact with the vehicle interior. The requirements are defined in FMVSS 201 [7-18] and SAE J 921 [7-19].

7.6.1.2 Torso Impact

To check the steering wheel and assembly unit, a body of 36 kg (79 lb) in accordance with SAE 944 [7-20] is used. The impact speed is 24 km/h (15 mph), and the test device represents a torso of a 50% male, as shown in Figure 7.13. Although the legal requirements are somewhat higher, most automobile manufacturers are designing their cars with values below 8000 N.

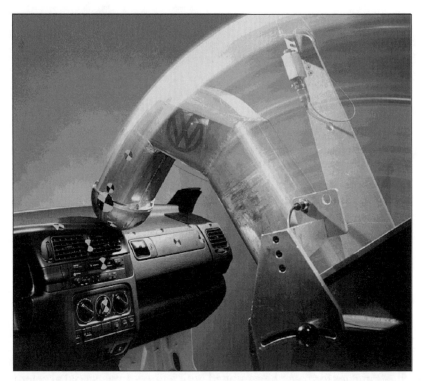

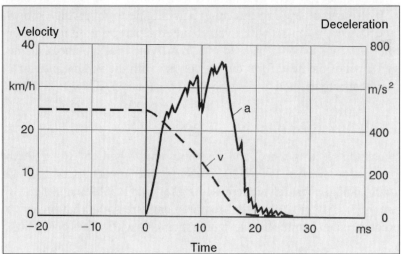

Figure 7.12 Head impact test device.

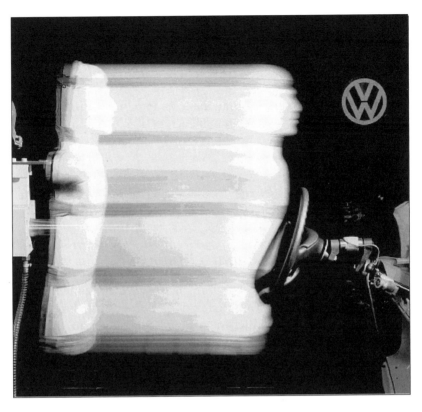

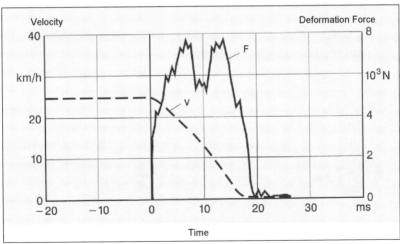

Figure 7.13 Steering column test

7.6.1.3 Pedestrian Accident Simulation Tests

The safety of pedestrians is a key issue in several regions of the world. In discussions, the following test devices are proposed:

- Child head form, 2.5 kg, 130-mm diameter (5.5 lb, 5.1-in. diameter)
- Adult head form, 4.8 kg, 165-mm diameter (10.5 lb, 6.5-in. diameter)
- Hip impactor
- Leg impactor

As shown in Figure 7.11, these devices should be used to evaluate the performance of cars having frontal impacts with pedestrians.

7.6.2 *Three-Dimensional Dummies*

The simulation of humans to test performance in accident simulation tests is a relatively new discipline. Until the mid-1960s, occupant simulation was done by sewn leather bags filled with sand. Since then, after equipping vehicles with sophisticated safety features, test dummies have been designed to more closely simulate human data, such as age, sex, and size. The measurement technique required today also demands a much more complicated design of dummy, as shown in Figure 7.14. Figure 7.15 shows the actual Hybrid III dummy.

In general, we also must take into consideration for the simulation of humans by dummies the following parameters:

- Degree of biofidelity, sensitivity to injury parameter
- Repeatability
- Reproducibility
- Durability
- Repairability
- Calibration technique
- Total cost

We likewise must differentiate, depending on the kind of investigation.

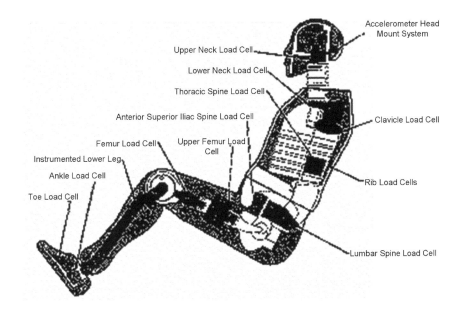

Accelerometer Head
Mount System

Upper Neck Load Cell

Lower Neck Load Cell

Thoracic Spine Load Cell

Anterior Superior Iliac Spine Load Cell

Clavicle Load Cell

Femur Load Cell

Upper Femur Load Cell

Instrumented Lower Leg

Ankle Load Cell

Toe Load Cell

Rib Load Cells

Lumbar Spine Load Cell

Figure 7.14 Hybrid III dummy, measurement possibilities.

7.6.2.1 Frontal/Rear Collision and Rollover

For the application mentioned previously, the 5% female, 50% male Hybrid III, and 95% male dummies are used. Figure 7.18 shows the dimensions of several dummies. The dummies must undergo a relatively complicated calibration test series before they are used in compliance tests. These dummies also are very expensive. In 2001, the cost of a 50% male dummy without instrumentation was approximately $30,000.

7.6.2.2 Lateral Impact

For lateral impact, two different 50% male dummies are in use—one for the United States, and one for European regulations. These are the U.S. side impact dummy (US-SID) and the European side impact dummy (Euro-SID). Figure 7.16 shows the differences between the two dummies [7-21].

The differences generally would not to be too big of a problem. However, because an identical car gives significant differences in the same accident,

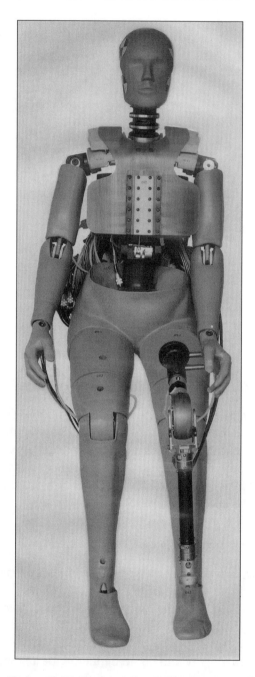

Figure 7.15 Design of a Hybrid III dummy.

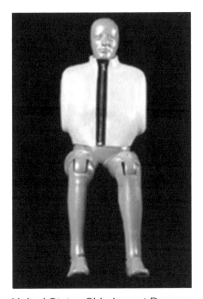
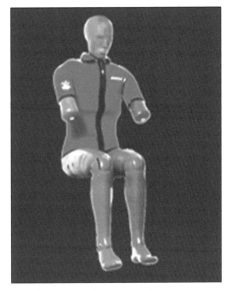

United States Side Impact Dummy,
US-SID

European Side Impact Dummy,
EuroSID-1

Figure 7.16 Comparison of US-SID and Euro-SID side impact dummies.

simulation tests make product development complicated if a car manufacturer is delivering its car to Europe and the United States. Design features that are positive for one side impact dummy appear to be negative for the other. In the future, as mentioned, there is some hope that we can use only one side impact dummy worldwide. Figure 7.17 shows the status of the year 2002 for the World-SID design [7-22].

7.6.2.3 Child Dummies

Child dummies are a special case that encompasses the simulation of babies up to 10-year-old children. Figure 7.18 shows some examples, in addition to the total range of child and adult dummies [7-22]. Child dummies are becoming increasingly important, not only for the evaluation of child restraints but in connection with the performance of front airbags.

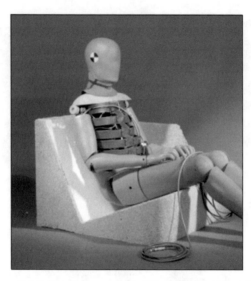

*Figure 7.17 Prototype of the World-SID, a worldwide
harmonized side impact dummy.*

7.6.2.4 Rear Impact Dummy (Bio-RID)

Because of the large number of rear-end collisions in Sweden, a new anthro-
pomorphic test device with special attention to the evaluation of possible
neck injuries in rear-end collisions was developed by the Denton company
with Chalmers University in Göteborg, Sweden. The Bio-RID dummy represents a 50% male with an articulated thoraco-lumbar spine and neck made
from a composite material. The motion of the cervical vertebra is controlled
by cables that are attached to neck-muscle substitutes and dampers. Details
are shown in Figure 7.19 [7-22].

7.6.2.5 Biofidelity Dummy

It also is often discussed whether the automobile industry needs a dummy
that more closely simulates human behavior. The answer to this question
depends on the purpose of the research analysis. For single research investigations, this might be acceptable. However, for general use in development
and certification work, it certainly is not.

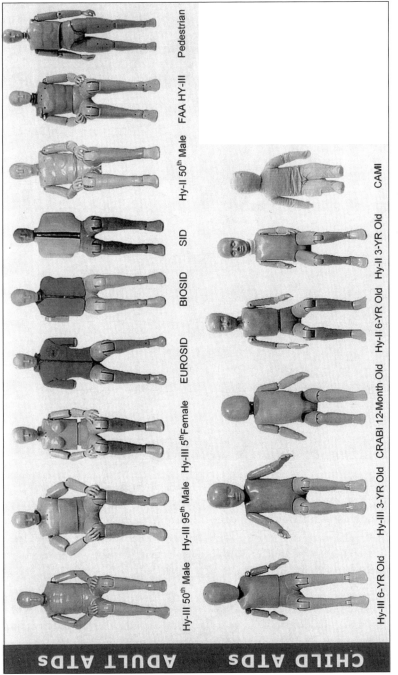

Figure 7.18 Dummy family. (Source: Ref. 7-22.)

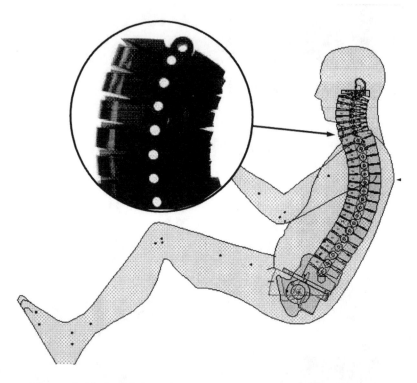

Figure 7.19 Bio-RID 2 dummy for rear impact applications.

7.6.3 Human and Dummy Modeling

For several applications, human models and parts of the human body are available. The models are used for accident simulation to reconstruct real-world accidents, mathematical optimization for the design of the vehicle, and biomechanical studies. Until now for certifications tests, the mathematical simulations have not been used, although the results meanwhile are as good as real-world accident simulation tests.

The simulation tools range from very simple models (one to two masses) up to very sophisticated tools. The most frequently used multi-body crash dummies are based on the MADYMO program. Figure 7.20 shows the wide application field [7-1, 7-23].

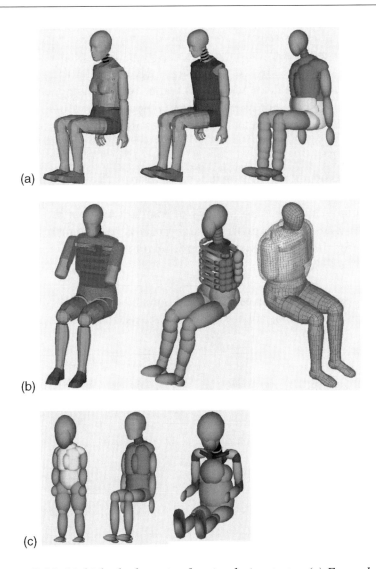

Figure 7.20 Multi-body dummies for simulation tests. (a) Examples of MADYMO Hybrid III dummy family (left to right): 5th percentile facet model, 50th percentile facet model, and 95th percentile ellipsoid model. (b) Examples of MADYMO side impact dummies (left to right): EUROSID-1 FE model, SID-IIs ellipsoid model, and US-DOT SID FE model. (c) Examples of MADYMO child dummies (left to right): CRABI 12-month ellipsoid model, Hybrid II six-year-old ellipsoid model, and TNO Q3 ellipsoid model.

MADYMO products range from a child dummy up to a 95% male adult dummy. The multi-body dummy gives very good results, especially for frontal impacts and using the restraint systems developed for this type of accident. Because of the detailed analysis that is available and the fact that the design engineer does not like to change computer software, finite element methods (FEM) simulation tools for dummies also are used [7-24], as shown in Figure 7.21.

The FEM computer model simulations compare well with the real dummy testing, especially in side impacts or in other simulation situations with a local dummy loading. Both types of simulation tools are used. The MADYMO system also gives good conformity to kinematics and to the influence of the restraint system, and it requires less computer time compared to the FEM system.

For more sophisticated investigations, a human body model is very expensive. Application fields are new vehicle systems that are not in production;

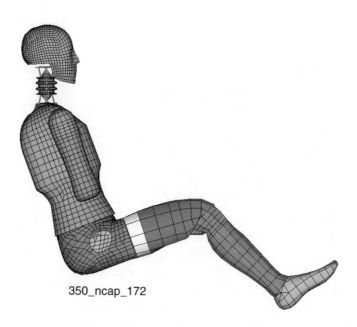

350_ncap_172

Figure 7.21 Finite element dummy.

they can simulate human parts such as the chest, neck, head, and feet. The simulation model therefore is complicated because of the behavior of humans. Geometric design data from Ramsis were used to design a series of human models of different sizes. Figure 7.22 shows multi-body human models representing a midsize male (left) and small female (right).

Vehicle manufacturers and large supplier companies use these human models for many of their analyses. Typical questions arise in areas where legislation is not available, but where the development engineer needs some support in the safety design.

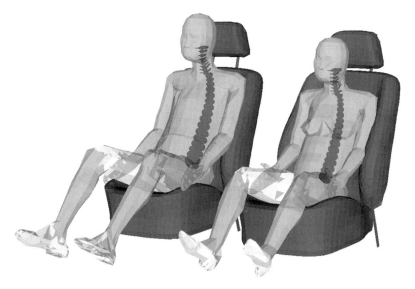

Figure 7.22 Simulation tools generated by the Ramsis model. Left: midsize male. Right: small female. (Source: Ref. 7-23.)

7.7 References

7-1. Wisman, J.S.H.M. et al. "Injury Biomechanics" (4 J 610), Proceedings of the Conference on Biomechanics, 3rd ed., Technical University of Eindhoven, The Netherlands, 2000.

7-2. Proceedings of the 8th Stapp Car Crash Conference, Detroit, 1964.

7-3. "The Abbreviated Injury Scale—1990 revision," Association for the Advancement of Automotive Medicine, Des Plaines, IL, United States.

7-4. Pike, J. *Automotive Safety*, Society of Automotive Engineers, Warrendale, PA, United States, 1990.

7-5. Swaeringen, J.J. "Tolerance of the Human Face to Crash Impact," Die Widerstandsfähigkeit des menschlichen Gesichtes gegen Stöße bei Unfällen, Federal Aviation Agency, July 1965, Report No. AM65-20, Oklahoma City, OK, United States.

7-6. Patrick, L., "Human Tolerance to Impact—Basis for Safety Design," SAE Paper No. 650171, Society of Automotive Engineers, Warrendale, PA.

7-7. Fiala, E., et al. "Verletzungsmechanik der Halswirbelsäule, Forschungsbericht der Technischen Universität, Berlin 1970.

7-8. Mertz, J.M., et al. "Strength and Response of the Human Neck," Proceedings of the 15th Stapp Car Crash Conference, Coronado, CA, 1971.

7-9. SAE Information Report. "Human Tolerance to Impact Conditions as Related to Motor Vehicle Design," Society of Automotive Engineers, Warrendale, PA, United States, 1986.

7-10. Kroell, C.K. "Thoracic Response to Blunt Fronted Loading," in *The Human Thorax, Anatomy, Injury, and Biomechanics*, P-67, Society of Automotive Engineers, Warrendale, PA, United States, 1976.

7-11. Stalnaker, R.L., Roberts, V.L., and Mc Elhaney, J.H. "Side Impact Tolerances to Blunt Trauma," 17th Stapp Car Crash Conference, pp. 377–408, Society of Automotive Engineers, New York, United States, 1973.

7-12. Viano, D.C. "A Viscous Tolerance Criteria for Soft Tissue Injury Assessment," *J. Biomechanics*, 2115, 1988, pp. 387–399.

7-13. U.S. Department of Transportation, National Highway Traffic Safety Administration, (NHTSA) and Federal Motor Vehicle Safety Standard (FMVSS), FMVSS 208, 49 CFR Parts 552, 571, 585, and 595 Occupant Crash Protection, Washington, DC, United States.

7-14. European Parliament and Council on the protection of occupants of motor vehicles in the event of a frontal impact and amending Directive 70/156/EEC, Brussels.

7-15. European Parliament and Council on the protection of occupants of motor vehicles in the event of lateral impacts and amending Directive 70/156/EEC, Brussels.

7-16. European Parliament and Council draft relating to the protection of pedestrians and other road users in the event of a collision with a motor vehicle, Brussels.

7-17. Friesen, F., et al. "Optimierung von Fahrzeugen hinsichtlich des Beinaufpralltests," *ATZ,* May 2002, Verlag Vieweg, Wiesbaden.

7-18. National Highway Traffic Safety Administration (NHTSA) and Federal Motor Vehicle Safety Standard (FMVSS), FMVSS 201, Occupant Protection in Interior Impact, National Highway Traffic Safety Administration, Washington, DC, United States.

7-19. U.S. Department of Transportation, National Highway Traffic Safety Administration. Laboratory Test Procedure for FMVSS 201 "Occupant Protection in Interior Impact Upper Interior Head Impact Protection," Washington, DC, United States. Also Test Procedure in Accordance with SAE J921 "Motor Vehicle Instrument Panel Laboratory Impact Test Procedure," Society of Automotive Engineers, Warrendale, PA, United States.

7-20. Test Procedure in Accordance with SAE 944a, Society of Automotive Engineers, Warrendale, PA, United States.

7-21. First Technology Safety Systems, Inc. Product description, Plymouth, United States, www.ftss.com.

7-22. Denton ATD, Inc. Product description, Milan, United States, www.dentonatd.com.

7-23. TNO Automotive. Description of MADYMO human models, Delft, The Netherlands.

7-24. Meywerk, M., et al. "Aspects of Optimization and Parameter Studies: Case Studies from Automotive Industry," LMS-Conference for Physical and Virtual Prototyping, Paris, France, 2001.

8.

Vehicle Body

8.1 General

The body-in-white and vehicle interior features contribute the most to vehicle safety, especially in the areas of reduction of low-speed damage and to the occupants of the vehicle in an accident. The body-in-white of a vehicle is not primarily designed to mitigate injuries but to carry the vehicle components (e.g., the powertrain and chassis) and the occupants of the vehicle. The first patent in the field of vehicle body design related to safety was published in October 1952 [8-1] by Bela Barény. In his patent, he described how the structural strength should be greatest in the vehicle compartment and that the front and rear of the vehicle should be less resistant to crushing and be capable of absorbing energy during a crash. Figure 8.1 shows the differences. The left side of the figure is a conventional design of that time; on the right side, the different stiffnesses in the front and rear end of the vehicle are compared to that of the passenger compartment.

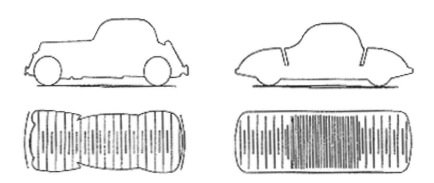

Figure 8.1 Patent drawing by Bela Barény.
(Source: Ref. 8-1.)

With a weight of approximately 250 kg (551 lb) in an average subcompact car design, the body-in-white absorbs slightly less than one-quarter of the total vehicle weight. Realizing the increased vehicle requirements that have occurred over the years, this low weight is a big improvement compared to the past. The car occupants gained more interior length for room and other advantages such as lower vibration and higher body stiffness of the vehicle. Figure 8.2 shows the increase in torsional stiffness from one model generation to the other for a compact car. Compact cars also have achieved a longer life durability and low corrosion over a lifetime of more than 10 years.

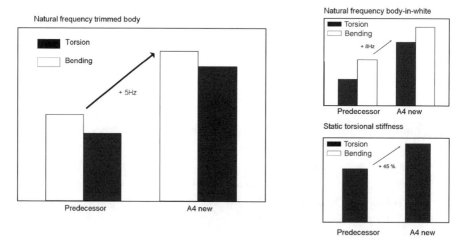

Figure 8.2 Comparison of the stiffness of a compact car for two generations of cars. (Source: Audi AG.)

With respect to accident distribution, Figure 8.3 lists information for European vehicles in which vehicle exterior body parts are involved in crashes [8-2].

Independent of the fact that in most accidents, the front of the vehicle is involved, all areas of the vehicle body could be involved in accidents.

The design of the body-in-white has changed during the last few years. Because of the necessity to minimize an increase in weight, regardless of the new requirements, each single part of the body-in-white was analyzed for

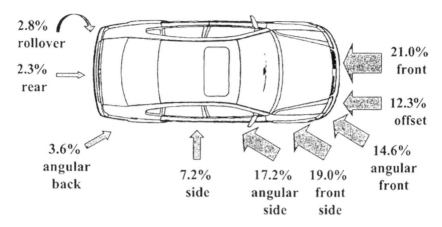

Figure 8.3 Statistical analysis of distribution of car accidents in Europe.

weight reduction and increases in performance. The following examples show different and successful designs. The Volkswagen Beetle as built until 1974 had a rigid underbody with a stiff front end, which also carried the front axle. Especially in car-to-car crashes, the design provided a good survival space, despite the low weight of the Beetle. Figure 8.4 illustrates the design.

Most new cars use the self-carrying body-in-white design. As an example of vehicles with a high production volume, the basic layout of the 2002 Opel Vectra [8-3] is shown in Figure 8.5.

Via a stable front cross-bar behind the bumper, the two longitudinal beams and the upper fender transmit the forcees in a crash to the middle tunnel and the A-pillar at the height of the floor panel. The doors, including the cross-bar for side impact protection, transmit forces to the rear of the vehicle. For the optimization of the longitudinal beams, many design parameters can be used. In addition to the optimum geometric configuration of the previously mentioned reinforcement at the place of the highest bending moment, some cars use longitudinal beams that are tailored metal blanks of different thick-nesses to ensure that the beams do not bend but fold. In the frontal area, these beams have a thickness of 1.5 mm (0.06 in.); the thickness in the rear is 2.5 mm (0.1 in.). To achieve a high energy absorption deformation, elements

Figure 8.4 Front end of the Volkswagen Beetle as of 1973.
(Source: Volkswagen Safety Report, 1974.)

with a high-volume specific energy-absorption capability are required. The force time function should be kept as constant as possible. Researchers have performed many studies in the field of collapse bulging. Recent investigations found that a "turn deformation" is even better.

The fuel tank must be placed in a well-protected area. Side and rear-end impacts should not disturb the fuel tank system nor the interior integrity of the passenger compartment. For rollover protection, the A- and B-pillars are reinforced. If the occupants of the vehicle are wearing seat belts, these pillars provide a good chance of survival in most rollover accidents.

Another interesting design is the DaimlerChrysler A-Class car [8-4]. Figure 8.6 shows the basic layout of this vehicle.

Figure 8.5 Body structure of a compact car. (Source: Ref. 8-3.)

Figure 8.6 The structural design of the DaimlerChrysler A-Class car.

Although the A-Class design fulfills the basic requirements in its class, the car is different because the front longitudinal beams are horizontal and are designed without a geometric "S" shape. Because bending beams are not absorbing enough energy, this alternative is one positive design feature. The other feature is the capability that in a frontal crash, the engine and transmission should glide under the underbody, thus allowing a higher crash length with respect to free deformation during frontal collisions. It also is one feature for a positive design with respect to compatibility in car-to-car crashes because the engine and transmission mass may not be harmful to the other colliding vehicle in some specific types of accidents.

Another interesting design is the aluminum space frame technology of the Audi A8 and A2 [8-5]. The design and production technique for the A2 is more advanced compared to the A8 of the first generation. More welded aluminum structures are used instead of cast knots. Figure 8.7 shows the specific design. Both vehicles could optimize the crash capability due to the space frame technology. Other techniques are used for the Aston Martin Vanquish, where many body parts are glued together.

Figure 8.7 The aluminum space frame of the Audi A2.

For convertible and roadster vehicles, additional features are necessary. One example used for the Mercedes SL is described as follows [8-6]:

- High occupant cell with multi-layer beams

- Composite body floor with optimized material thickness including diagonal, longitudinal, and transverse beams

- Increase in the use of high-strength steel from 19% (predecessor) to 33%

- Optimized wheel

- Installation of a multi-piece elliptical front wall with a cross member

- Additional beams in the upper front structure

- Bending-resistant B-pillar with lateral longitudinal beams

- Doors with reinforcement

- Integral seat with stiff cushion seat-back frame

- Reinforced A-pillar

- Automatic rollover bar

Many other available designs are on the road, such as sport utility vehicles (SUVs), multi-material concepts used in some sports cars, and one-box designs for multi-purpose vehicles. However, these fit more or less into the previously demonstrated concepts.

As mentioned, the body-in-white has many more duties. Figure 8.8 demonstrates the numerous different forces that must be taken into consideration by a good design over several 100,000-km (62,305-mi) driving distances and more than 10 years of lifetime.

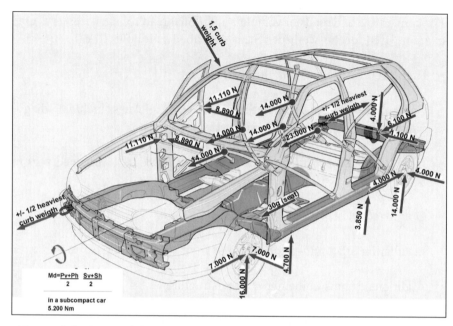

Figure 8.8 External forces to the body-in-white. Note: Numerical values in German spelling (7.000 N = 7,000 N in American spelling).

8.2 Low-Speed Impact

Several requirements are defined for the testing of vehicle integrity at low-speed impacts and the performance of the vehicle in relation to repair costs at high-impact speeds. Figure 8.9 provides an overview of the requirements in the front of the vehicle, and Figure 8.10 provides an overview of the requirements for the rear of the vehicle.

The results are published in the consumer information systems and by insurance companies. These results are used as the basis for vehicle damage assessment of the relevant vehicle, as well as for the basis for insurance costs [8-7].

Figure 8.11 shows a good example of a positive design in low-speed impacts.

Related to repair costs at high speeds are not only the repair methods and associated costs but the costs of the spare parts that also influence the rating.

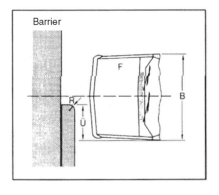

Offset crash 40% overlap
Ü[overlap] = 0.4 · B
B is the width of the vehicle without exterior mirrors
Impact steering wheel side
v_F = 15 kph + 1 kph
Barrier clearly higher than the front of the vehicle
R = 150 mm
No contact between the front of the vehicle and the wall next to the barrier
Vehicle occupied by one dummy, 50th percentile male, driver's position,
   restrained
Fuel tank full of gasoline or diesel, water permissible
Suspension alignment check before and after crash test
Vehicle driveable and coasting before impact

Figure 8.9 Repair cost test (front of the vehicle).

In addition to the design, the vehicle manufacturer has other possibilities (e.g., low prices on spare parts) to reduce insurance premiums.

8.3 Vehicle Body Test Without the Simulation of Car Occupants

In the real-world crash environment, the vehicle occupant plays the major role in combination with the vehicle related to occupant protection. However, many tests involve the vehicle body without the use of a three-dimensional dummy.

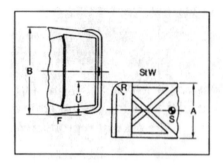 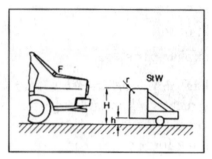

Offset crash 40% overlap
Ü overlap = 0.4 · B
B is the width of the vehicle without exterior mirrors
Impact on steering wheel side
StW is the moving barrier, coasting before impact
mStW = 1000 kg
Wheelbase $d_r \geq 1.5$ m
A is the width of the moving barrier ≥ 1.2 m
S is the center of gravity of the barrier in the center plant
H is the barrier height 700 mm
H is the lower edge barrier 20 mm
R = 150 mm
r = 50 mm
vStW = 15 kph + 1 kph
v_F = 0 kph, brakes applied
No contact between the front of the vehicle and the wall next to the barrier
Vehicle occupied by one dummy, 50th percentile male, driver's position, belted
Fuel tank filled with gasoline or diesel, water permissible
Suspension alignment check before and after crash test
Vehicle ready to drive

Figure 8.10 Repair cost test (rear of the vehicle).

8.3.1 Quasi-Static Test Requirement

8.3.1.1 Seat and Seat-Belt Anchorage Point Tests

If the lower inner anchorage point is mounted at the seat, which usually is the case for the seat-belt latch, the seat and the seat-belt anchorage points are tested simultaneously. Figure 8.12 shows a typical test configuration.

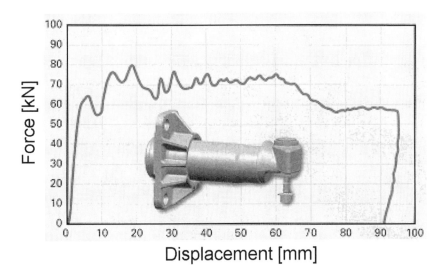

Figure 8.11 Design feature for good performance in low-speed impacts (repair cost). (Source: Ref. 8-8.)

By using rigid body blocks, the pulling force is applied equally to each anchorage point. (The test is done in accordance with FMVSS 210 [8-9].) For each occupant seat, the resisting force should be greater than 14,000 N. The upper anchorage point with the variable height adjustment and the latch at the seat require special attention. For example, the reinforcement plate at the B-pillar should not be stiff to prevent the outer B-pillar metal from being cut. The belt latch at the seat also is important because the forces are transmitted via movable metal parts to the stiffer portion of the underbody or the vehicle middle tunnel. The reason for this is that in a standard seat design, the seat cannot absorb the high forces from the seat-belt anchorage points. The seat itself must resist more than 20g over a period of more than 30 ms. To transmit the force to the middle tunnel, a serrated seat rail is used. Because of the loading during an accident, the vertical component of the seat-belt pulling force snapped the seat-belt anchorage point firmly into this special design element. Figure 8.13 shows the design of such a solution (serrated strip).

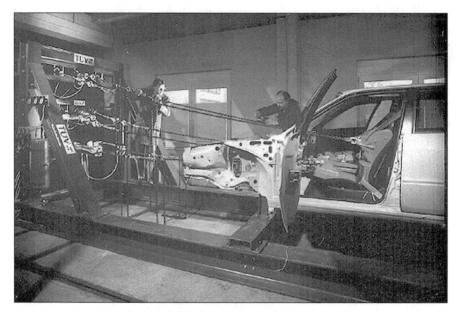

Figure 8.12 Seat-belt anchorage test.

Figure 8.13 Design of the seat track mount and the seat track.

Seats in which the upper anchorage point also is mounted at the seat back must have reinforcements installed in the seat back to absorb the forces and moments. This often requires more weight and costs, and thus such a solution is installed only in special models such as convertibles.

8.3.1.2 Roof Strength

For the evaluation of roof strength in accordance with FMVSS 216 [8-10], a steel plate is used. This plate is inclined by 25° in relation to the horizontal vehicle longitudinal plane and 5° to the front. Figure 8.14 shows the test configuration.

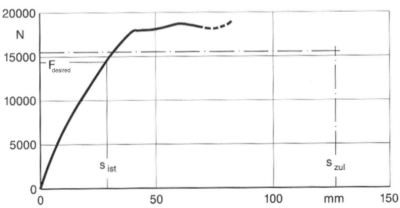

Figure 8.14 Roof test.

The legal requirement requests that with a force that is 1.5 times the vehicle curb weight (but not greater than 2267 kg [4998 lb]), the deformation measured perpendicular to the test plate should not exceed 12.7 cm (5 in.). This requirement also must be met in convertibles. Because of the test configuration, the greatest resistance is created by a reinforced A-pillar, which together with the glued-in windshield and the roof can provide sufficient survival space in case of a rollover accident.

8.3.1.3 Vehicle Side Structure

In addition to the dynamic tests, the side structure of the vehicle is tested in a quasi-static test, in accordance with FMVSS 214 [8-11]. This test uses a half cylinder that is pushed perpendicular to the longitudinal axis of the vehicle into the door of the tested vehicle. The lower part of the cylinder has a height of 12.7 cm (5 in.) above the lower part of the door sill. The cylinder has diameter of 30.5 cm (12 in.). The height is set so that the upper part of the cylinder is at least 12.7 cm (5 in.) higher than the lowest part of the bottom edge of the side windows. Figure 8.15 shows the test arrangement and the intrusion force as a function of deformation length of the door.

The lower curve describes the result for a door without a side beam, whereas the upper curve describes the result for a door with a side beam. During such tests, the reachable maximum force is limited by the force transmission of the door, including the side beam, to the hinges and the door latch. In some new designs, the lower part of the door also is anchored with the door sill. Because the maximum force is limited by the resistance of the A- and B-pillars and the lower anchorage mechanism, the maximum force is approximately the same for the door without the side beam. The real positive effect occurs during the first 240 mm (9.45 in.) of deformation. This means that especially at the beginning of an impact with another car, the resistance on the side of the impacted vehicle should be as high as possible.

Some research engineers believe that the side of the vehicle should be built similar to a board frame, as shown in Figure 8.16 [8-12]. If the side of the struck vehicle and the front of the impacting vehicle are designed to be compatible, a slide-away effect could reduce the potential of intrusion and the risk of injuries.

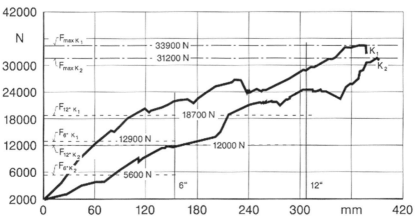

Figure 8.15 Quasi-static lateral test.

Because of new dynamic tests with modern dummies, the question generally arises from time to time whether these quasi-static tests are of value in improving safety. Although the tests are far from real-world accident performance, we believe that these minimum requirements are a good starting point for the designer to lay out the vehicle structure.

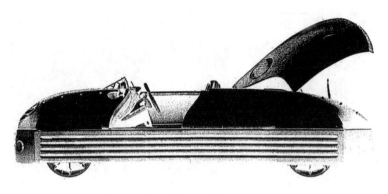

Figure 8.16 Board frame layout.

8.4 References

8-1. Barényi, Bela. Kraftfahrzeug, insbesondere zur Beförderung von Personen, Deutsches Patentamt Nr. 854157 30, October 1952.

8-2. Fürstenberg, K.Ch., et al. "Development of a Pre-Crash Sensorial System—The CHAMELEON Project," in VDI-Berichte 1653, Vehicle Concepts for the 2nd Century of Automotive Technology, ed. by Verein Deutscher Ingenieure, VDI-Verlag, Düsseldorf, Germany, 2001, pp. 289–310.

8-3. Zimmermann K.-H. "Der Neue Opel Vectra," *ATZ*, February 2002, Vieweg-Verlag, Wiesbaden, Germany, ISSN 0001–2785.

8-4. Jung C., et al. "Rohbau der Karosserie und Passive Sicherheit," *ATZ/MTZ Special Edition*, 2001, Vieweg-Verlag, Wiesbaden, Germany.

8-5. "Der Neue AUDI A2, *ATZ/MTZ*, *Special Edition*, March 2000, Verlag Vieweg, ISSN 0001-2785.

8-6. "Mercedes SL 2001," *ATZ/MTZ Extra*, October 2001, Vieweg-Verlag.

8-7. Anselm, D. *The Passenger Car Body*, Society of Automotive Engineers, Warrendale, PA, 2000.

8-8.   Haberer, K.H. "Das Stoßfängersystem des AUDI A4, Der Neue AUDI A4," *ATZ Special Edition*, Vieweg-Verlag, Wiesbaden, November 2000.

8-9.   National Highway Traffic Safety Administration (NHTSA) and Federal Motor Vehicle Safety Standard (FMVSS), FMVSS 210, Seat Belt Assembly Anchorages, National Highway Traffic Safety Administration, Washington, DC, United States.

8-10. National Highway Traffic Safety Administration (NHTSA) and Federal Motor Vehicle Safety Standard (FMVSS), FMVSS 216, Roof Crush Resistance, National Highway Traffic Safety Administration, Washington, DC, United States.

8-11. National Highway Traffic Safety Administration (NHTSA) and Federal Motor Vehicle Safety Standard (FMVSS), FMVSS 214, Side Impact Protection, National Highway Traffic Safety Administration, Washington, DC, United States.

8-12. Schimmelpfennig, K.-H. Unpublished data, Münster, Germany.

9.
Dynamic Vehicle Simulation Tests

9.1 Frontal Collisions

The frontal collision was the first type of accident that was analyzed in detail for occupant protection. In addition to the previously mentioned low-speed tests, impact speeds of up to 64 km/h (40 mph) are being used today. The obstacles vary from a rigid fixed wall, another vehicle, a pole, and a tree simulation to various deformable barriers. The basic elements of a frontal collision should be explained by an analysis of a frontal collision against a fixed barrier. The kinetic energy must be absorbed by the vehicle and by some elastic rebound. Figure 9.1 shows the deceleration $\ddot{s}$, the velocity change $\dot{s}$, and the deformation length s as a function of time.

The rebound, visible by the negative velocity, shows that for this type of accident simulation, the elastic rate is approximately 10%. This means that the change in velocity for an impact against this fixed barrier at 50 km/h (31 mph) is approximately 55 km/h (34 mph). For a quick analysis, we can assume a non-elastic impact performance. The formulas describe the following:

$\ddot{s}_v$ = Deceleration of the vehicle as function of time f(t)

$\dot{s}_v$ = Velocity of the vehicle during the impact as f(t)

s_v = Deformation of the vehicle during the impact as f(t)

F, $\overline{F}$ = Deformation force, average deformation force

v_i = Impact velocity

Δv = Change in velocity

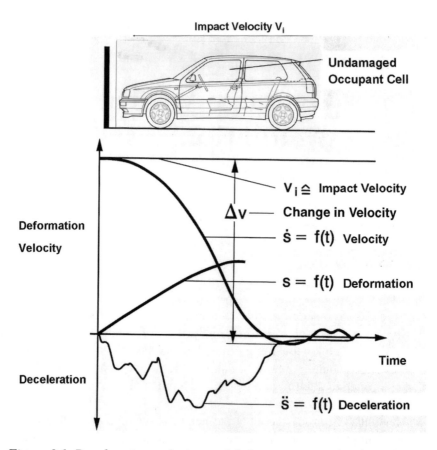

Figure 9.1 Deceleration, velocity, and deformation as a function of time.

If we further stipulate that the average deformation force remains constant over time, then the following correlation is valid:

$$\ddot{s}_v = -a, \; \dot{s}_v(t=0) = v_i, \; s_v(t=0) = 0$$

$$\ddot{s}_{v(t)} = -a \quad \text{with} \quad \frac{m_v}{2}\left(v_i^2 - \dot{s}^2\right) = \int_0^s F \cdot d_s = \overline{F} \cdot s_v$$

$$\dot{s}_{v(t)} = -a \cdot t + v_i$$

$$s_{v(t)} = \frac{-a \cdot t^2}{2} + v_i t$$

$$s_{v(s)} = \frac{v_i^2 - \dot{s}^2}{2a}$$

If we also replace the average deformation force through the expression $\overline{F} = m_v \cdot a$, then we find another interesting relationship:

$$\frac{m_v}{2} \cdot v_i^2 = m_v \cdot a \cdot s_v$$

or

$$v_i^2 = 2a \cdot s_v$$

From that formula, we can conclude that the deformation force could be different from the car-to-car design. As vehicle mass increases, so does the force, $\overline{F} = m \cdot a$. If the average deceleration in frontal barrier impacts is similar among the cars on the road, then the deformation forces are mass dependent. In reality, smaller cars have a shorter deformation length and thus have a higher g-level during frontal impacts. The deformation length for cars at an impact against a fixed barrier at 50 km/h (31 mph) varies between 450 to 750 mm (18 to 29.5 in.). At 56 km/h (35 mph), it reaches values of 550 to 850 mm (22 to 33.5 in.). Special cars such as micro-compact vehicles (i.e., smart electric vehicles) have smaller deformations, in the range of 300 of 350 mm (12 to 14 in.). Although no measurable lateral force component occurs during a straight, frontal collision, we find many transverse forces in other crash modes.

In addition to the acceleration measurement on the tested vehicle, some test facilities measure the reaction force at the barrier. Figure 9.2 shows the resultant force, measured as single forces with 72 segments in the front of the barrier and summed to the total force, versus time and deformation [9-1].

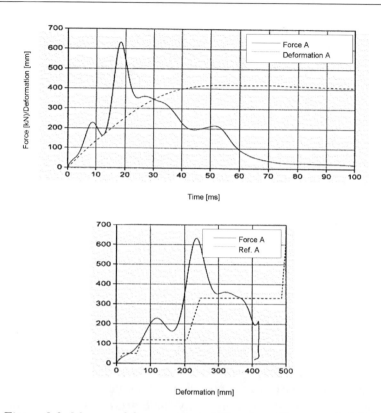

Figure 9.2 Measured forces at a rigid barrier. (Source: Ref. 9-1.)

Figure 9.3 shows a modern instrumented deformable crash barrier, which uses 64 independent triaxial load sensors [9-2].

The results from these measurements allow the following interpretation: Most forces are transmitted at the bumper height approximately 400 to 500 mm (16 to 20 in.) above the ground. This is not valid for vehicle types other than passenger cars. Relatively low forces are recorded from the bottom to 400 mm (16 in.) and above 700 mm (27.5 in.).

The type of barrier significantly influences the acceleration and deformation level of the tested vehicle. Although during the straight central impact against a fixed wall which is due only to the vehicle asymmetry, small lateral accelerations occur. Other crash modes clearly show visible deceleration in the x and y

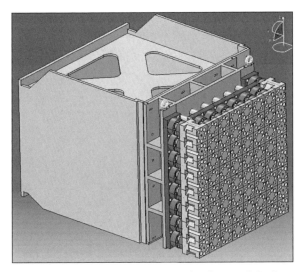

Figure 9.3 Measurement device at the front of the barrier.
(Source: Ref. 9-2.)

directions. Also, the shape of the velocity change versus time function is different. Figure 9.4 demonstrates the deceleration as a function of time for a frontal 90° crash, a 30° barrier, and an offset deformable barrier crash test. Also, the change in velocity is shown for the various crash types. Because the crash against the 30° barrier was performed only at 48 kph (30 mph), only the tendencies of the $\ddot{s}$, $\dot{s}$ curves as a function of time are relevant.

From the data measured, we can conclude that the deceleration level is highest for the frontal 90° crash. This also means the strongest requirements with respect to the performance of the restraint systems. The offset crash produces a higher loading on the vehicle structure because only one front longitudinal beam often is taking the highest load in relation to the energy absorption. Other frontal collision tests that are used by some vehicle manufacturers are discussed next.

9.1.1 Pole Test

With impact speeds up to 29 ± 0.5 km/h (18 ± 0.3 mph), the test is performed against a rigid pole that has a typical diameter of $\cong 250$ mm ($\cong 10$ in.).

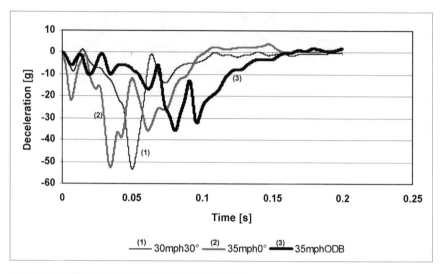

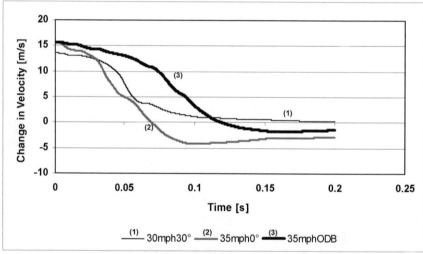

Figure 9.4 Deceleration and velocity time function for different types of crashes. (Source: Ref. 1-1.)

9.1.2 Frontal Car-to-Car Crash

For compatibility tests, see Chapter 13 for accident simulation tests, and for developmental reasons, vehicle manufacturers also are undertaking vehicle-to-vehicle tests. In a frontal collision between two vehicles with the masses m_1 and m_2, the following relationship is valid:

$$\frac{m_1 \cdot v_{i1}^2}{2} + \frac{m_2 \cdot v_{i2}^2}{2} = \frac{1}{2}(m_1 + m_2)u^2 + \overline{F} \cdot (\Delta s_1 + \Delta s_2)$$

where

v_1 is the impact speed
u is the common speed after the impact

$$u = \frac{m_1 \cdot v_{i1} + m_2 \cdot v_{i2}}{m_1 + m_2}$$

If we insert the relative velocity (v_r) at the start of the impact between both vehicles

$$v_r = v_{i1} - v_{i2}$$

the two preceding equations result in

$$v_r = 2\overline{F}(\Delta s_1 - \Delta s_2) \cdot \frac{m_1 + m_2}{m_1 \cdot m_2}$$

With this expression, the change in velocity for Vehicle 1 and Vehicle 2 is as follows:

$$\Delta v_1 = v_{i1} - u = (v_{i1} - v_{i2}) \cdot \frac{m_2}{m_1 + m_2} = v_r \cdot \frac{m_2}{m_1 + m_2}$$

and

$$\Delta v_2 = v_{i2} - u = (v_{i2} - v_{i1}) \cdot \frac{m_1}{m_1 + m_2} = v_r \cdot \frac{m_1}{m_1 + m_2}$$

It is interesting to see that the change in velocity is influenced only by the vehicle masses involved and the relative collision velocity. A frequently discussed question is: Which velocity must be chosen in a collision against a fixed wall to simulate a frontal collision of two identical vehicles against each other? If we define v_{i1} equal to v_{i2}, that means that each vehicle has the same but opposite speed. One condition is that the change in velocity Δv_{1B} must be the same as during the vehicle-to-vehicle and vehicle-to-barrier test. For a non-elastic impact, the following relationship is relevant:

Case 1: Impact Against a Fixed Barrier.

$$\Delta v_{1B} = (v_{i1} - v_{i2}) \cdot \frac{m_2}{m_1 + m_2}$$

With m_2 equal to the barrier with a mass ∞ and v_{i2} equal to the speed of the fixed barrier, this means zero. The change in velocity in the crash against a fixed barrier is $\Delta v_{1B} = v_{i1}$.

Case 2: Impact Against a Second Vehicle.

$$\Delta v_{1veh} = (v_{i1} - v_{i2}) \cdot \frac{m_2}{m_1 + m_2}$$

With m_2 equal to m_1 (identical vehicles) and v_{i1} equal to v_{i2} identical but opposite speed, the change in velocity in a vehicle-to-vehicle test becomes $\Delta v_{1veh} = v_{i1}$.

From this short analysis, we can see that a collision against a fixed barrier with v_{i1} has the same consequences with respect to the change in velocity, a non-elastic impact assumed, as a vehicle-to-vehicle collision, with each vehicle having the same mass, structure, and impact speed. To make this result more transparent, the following example is used: A vehicle impact with 50 km/h (31 mph) against a fixed barrier has the same consequences as an impact of two identical cars with the same mass with a relative impact speed of 100 km/h (62 mph). If we consider that for the production car we must take into account 10% elasticity, the 50 km/h (31 mph) barrier impact has a change in velocity of 55 km/h (34 mph), and the V_{rel} crash of 100 km/h (62 mph) between two cars becomes 110 km/h (68 mph).

This performance changes if the vehicle masses are different. Figure 9.5 shows the Δv and the deceleration as a function of collision time for two cars with $m_1 = 966$ kg (2130 lb) and $m_2 = 1960$ kg (4321 lb) ($m_1/m_2 = 1/2.03$). The relative velocity between both cars at the beginning of the impact was 28 m/sec (92 ft/sec).

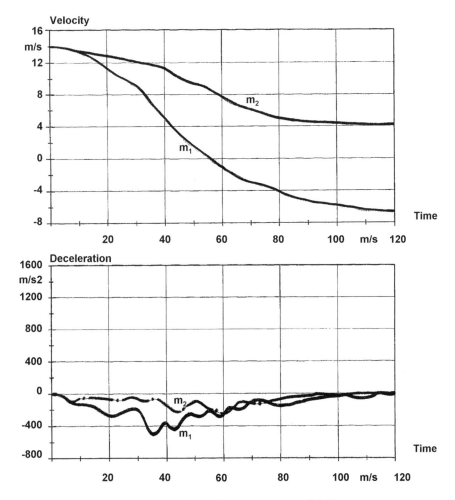

Figure 9.5 Frontal collision between cars of different masses.

In the relevant time period for the occupant protection, the Δv for the heavier vehicle is 10 m/sec (33 ft/sec) and for the lighter vehicle is 20.5 m/sec (67 ft/sec). This is very close to the results of the formula mentioned previously.

$$\Delta v_{ml} = 28 \cdot \frac{1960}{2926} = 18.76 \text{ m/sec}$$

$$\Delta v_{m2} = 28 \cdot \frac{960}{2926} = 9.25 \text{ m/sec}$$

9.1.3 Design Requirements of Frontal Collisions

Regardless of the occupant restraint system, some basic requirements must be fulfilled by the vehicle structure in frontal collisions. The longitudinal front bars must be designed to be as horizontal as possible. In a front-wheel-drive configuration, the driveshafts usually require an opening, which gives to the longitudinal beams a small upward tilt, followed by a downward configuration, the so-called S-shape. Therefore, the longitudinal beams in this area are reinforced by metal shields to avoid bending. The force from the beams is transmitted to the outer part of the floor panel and the middle tunnel of the vehicle. Also, the front transverse beam must be strong enough to not only resist the forces created in low-speed impacts but to engage both front longitudinal beams in offset crashes. The height of the longitudinal and transverse beams should be between 400 to 500 mm (16 to 20 in.) above the ground.

Although the upper part of the front end is not the major component with respect to energy management, it contributes to maintaining the survival space for the occupant. This is important not only for the function of the restraint system, but for the necessity to open one door without tools after the crash. The sidewalls must be firmly connected with the wheel housing, and the force in the middle of the A-pillar must be transmitted via the door and its reinforcement (e.g., the side beam, to the B-pillar).

Another important requirement is the integrity of the windshield. The laminated windshield must remain in its frame to act as a reacting part for the airbag. In accordance with FMVSS 212 [9-3], the front hood must not penetrate the windshield from the outside in a predefined zone. Figure 9.6 [9-4] provides a good estimate of how the energy absorption is distributed via the front end of a vehicle. If we sum these forces, approximately 50% of the energy is absorbed by the longitudinal beams.

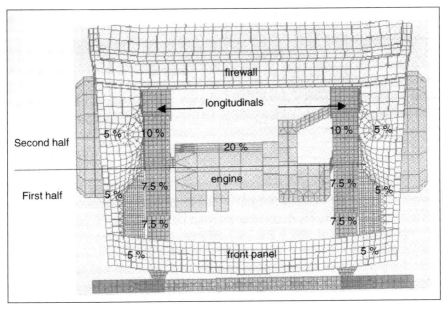

*Figure 9.6 Energy absorption of different vehicle parts.
(Source: Ref. 9-4.)*

In addition to the fact that the steering wheel requirement technically is over-ruled by FMVSS 208 [7-13], in accordance with FVMSS 204 and EEC 74/297/ECCR 12, the steering wheel must not have a relative movement to the body-in-white that exceeds a horizontal rearward movement and an upward movement of more than 127 mm (5 in.) during a 50-km/h (31-mph) frontal barrier impact. The design solutions for these requirements are stable cross-bars under the dashboard and steering column, which have a mechanism to collapse them during impact.

Because of the higher degree of protection, if the seat belts are used, and in combination with airbags, other injuries become more important. This is why special attention is given to the intrusion level of the foot pedals of the vehicle in frontal crashes. This intrusion also is one of the assessment criteria for the NCAP test and therefore is an important design criteria.

9.2 Lateral Collisions

According to recent accident statistics (see Figure 9.7), the lateral collision has become a high priority.

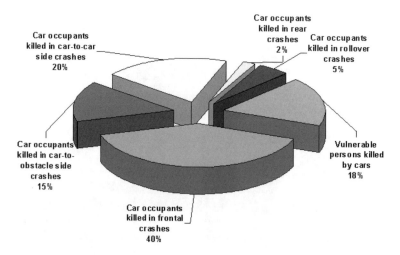

Figure 9.7 Distribution of fatal accidents. (Source: Ref. 9-5.)

The explanation for this is the greater degree of occupant protection in frontal impacts, which was achieved by design efforts made during the past few decades. There are basically three legal requirements for dynamic side impact testing:

- Impact of 4000 lb = 1818 kg rigid barrier with a speed of 32 km/h (20 mph) under 90° to the longitudinal axis of the tested car. This test is performed primarily to check the integrity of the fuel tank system during and after the test.

- Impact of a moving barrier in accordance with FMVS 214 [8-11]. This test has been the basis for legal requirements since 1993 in the United States. The barrier with a weight of 1365 kg (3000 lb) is crashed with $\cong 54$ km/h ($\cong 33.5$ mph) in a crabbed configuration, as shown in Figure 9.8.

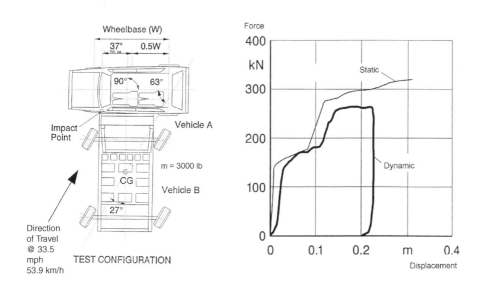

Figure 9.8 Test configuration in accordance with FMVSS 214.

The impact point is defined as follows: the left corner must have a distance of 0.5l + 0.94 m (1.7 + 0.6 ft) from the center of the rear wheel, where l is the wheelbase of the impacted vehicle centerline of the moving barrier.

On the right part of Figure 9.8, we also can see the test configuration and the force deflection characteristic of the deformable element, which is attached to the front of the barrier. Figure 9.9 shows the velocity as a function of time for the tested vehicle and the barrier, with the acceleration of the impacted vehicle and of the upper spine of the dummy.

- The third test is the impact of a movable/deformable barrier in accordance with EEC regulation No. 96/27/EEC [7-15]. This barrier is impacting the

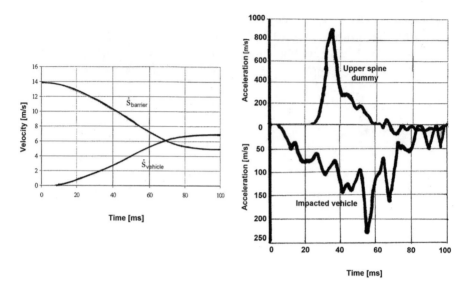

Figure 9.9   Acceleration, velocity as a function of time in a test in accordance with FMVSS 214 for vehicle and occupant.

tested vehicle under 90° with the centerline meeting the R-point. The details of the barrier are specified in the amendments to the previously mentioned rule. Of special importance is the layout of the impact device, which shows (depending on the location of the impact element) different force/deformation characteristics. The barrier has a weight of 950 kg (2094 lb), and the impact speed is 50 km/h (31 mph), as shown in Figure 9.10.

Figure 9.11 shows the differences of the impacting barriers as defined by the National Highway Traffic Safety Administration (NHTSA) and Economic Commission for Europe (ECE).

The EEVC barrier is higher above the ground and is smaller in width than that of the United States. Regardless of the differences in impact speed, the deformation of the tested vehicles also differs; that is, the ECE does not hit the body sill in most cases.

For the last two tests mentioned, it is important that the vehicle be designed in some areas especially to resist the impacting barrier. This means a strong

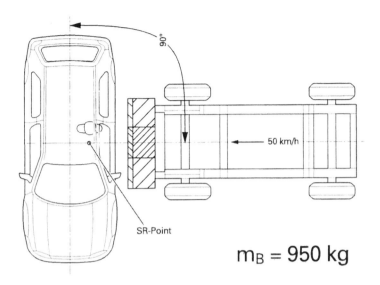

$$m_B = 950 \text{ kg}$$

*Figure 9.10 Test configuration in accordance
with EG 96/27 and ECE R 95.*

A+B-pillar, strong doors and hinges, reinforcements in the doors, and a good connection of the lower doors to the outside of the underbody with a strong cross member in the area of the seat mounting system (front and rear). Figure 9.12 shows the reinforcement measures for a subcompact vehicle for side impact protection [9-6].

9.3 Rear-End Collisions

For the rear-end collision, the same test is used as for one side impact test with the rigid movable barrier with a weight of 1800 kg (3968 lb) and a speed of more than 48.3 km/h (30 mph). The assessment criterion for the tested vehicle again is the integrity of the total fuel tank system. Figure 9.13 shows the acceleration and velocity as a function of time for the impacted vehicle.

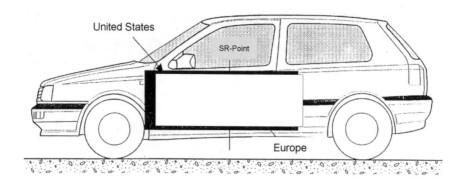

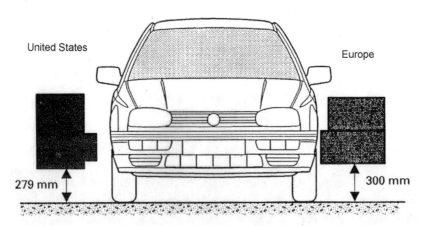

Figure 9.11 Differences of the side impact between barrier tests for the United States and Europe.

9.4 Rollover

For the rollover situation, two different kinds of tests are performed. The integrity of the fuel tank system is tested with the help of a test device, which can turn the vehicle step by step up to 360° (see Figure 9.14).

The vehicle is mounted in this device and, via 90° steps, is tested to determine whether the fuel tank and fuel lines show any signs of leakage. The dwell time in each position is at least 5 min. To prevent fuel loss, gravitation valves are installed between the ventilation pipe and the active charcoal filter.

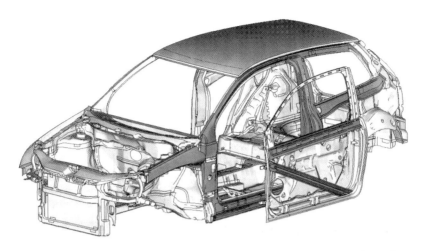

Figure 9.12 Crash-related structural reinforcements at the side of the vehicle and the door.

The second procedure is used in connection with the analysis of the performance of the vehicle occupants during accident simulation tests. The tested vehicle is put onto a moving sled under an angle of 23°, as shown in Figure 9.15.

The sled with the vehicle on top has a speed of 48.3 km/h (30 mph) and is stopped via a controlled deceleration. The vehicle then starts to roll several times, as shown in Figure 9.16

Compared with frontal and lateral impacts, the time sequence of the rollover simulation test is much longer. Whereas the frontal and side impacts are finished after 100 ms, the rollover is finished after several seconds—in the case shown, at approximately 4000 ms.

The body-in-white normally has good strength to resist these impact loads. For a two-seater vehicle, the A-pillar with the glued-in windshield and, very often, reinforced seat backs with sufficient height offer satisfactory protection to occupants during rollover accidents if the seat belts are worn by the occupants. For some convertibles, automated roll-bars, seat-back reinforcements, or vertical movement protection devices are used. These are activated

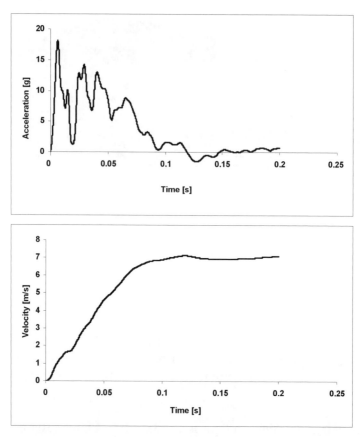

Figure 9.13 Rear-impact data in accordance with FMVSS 301.

by sensors. A typical design for such a protection device, directly behind the rear seat back, is shown in the 2002 Audi A4 convertible [9-7] (see Figure 9.17). Behind both seat backs of the rear passengers, two rollover protection systems are installed. In severe accidents (e.g., rollover or front, lateral, or rear-end collisions), a pre-loaded spring is released after the sensors, and the ECU determines the severity of the accident.

Figure 9.14 Test configuration for fuel tank integrity.

*Figure 9.15 Test configuration for the dynamic rollover test
in accordance with FMVSS 208.*

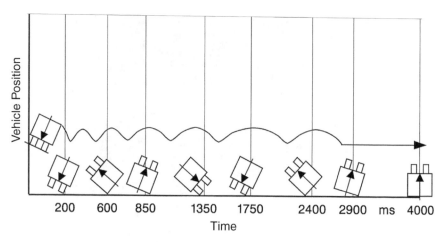

Figure 9.16 Vehicle behavior as a function of time during a dynamic rollover test. (Source: Ref. 1-1.)

Figure 9.17 Automated rollover protection in a convertible. (Source: Ref. 9-7.)

9.5 References

9-1. Relou, J. *Methoden zur Entwicklung crashkompatibler Fahrzeuge*, Shaker Verlag, 2000, Aachen, ISBN 3-8265-7804-x.

9-2. Compagnie BiA. Product information, Conflans Ste. Honorine, France, February 2002.

9-3. National Highway Traffic Safety Administration (NHTSA) and Federal Motor Vehicle Safety Standard (FMVSS), FMVSS 212, Windshield Mounting, National Highway Traffic Safety Administration, Washington, DC, United States.

9-4. Wittemann, W. *Improved Vehicle Crashworthiness Design by Control of the Energy Absorption for Different Collision Situations*, Technical University of Eindhoven, The Netherlands, 1999, ISBN 90-386-0880-2.

9-5. Klanner, W. "Status Report and Future Development of the EURO NECAP Program," Experimental Safety Vehicle (ESV) Conference 2001, Amsterdam, The Netherlands.

9-6. Heumann, G., et al. "Sicherheitskonzept des neuen Volkswagen Polo," Berichte des Aachener Kolloquiums Fahrzeug- und Motorentechnik, 2001.

9-7. Feldschmid, A., et al. "Das Neue Audi A4 Cabriolet," *ATZ/MTZ*, Vieweg-Verlag, Wiesbaden, March 2002.

10.

Occupant Protection

10.1 Vehicle Compartment

Aside from the classic restraint system, which consists of the combination of seat belts and airbags, the total passenger compartment must fulfill certain requirements. For example, the minimum head impact zone is defined by a sphere of 165 mm (6.5 in.) diameter and requires no sharp edges by a radius of 2.5 mm (0.1 in.) or 3.2 mm (0.126 in.), depending on the location. Another requirement is defined in the newer version of FMVSS 201. In this standard, it is requested that a head form with a speed of 24 km/h (15 mph) is impacting certain interior parts of the occupant compartment. In these impact zones, which also include the side head airbag, special deformation measures must be taken. Figure 10.1 shows one example of the impact test investigation using an FEM calculation.

To keep the acceleration level low, a deformation length in the direction of the impact of up to 50 mm (2 in.) is the basis to fulfill the specified level of less than 80g deceleration.

10.2 Restraint Systems

With respect to restraint systems, we must differentiate between devices that must be activated manually by the occupant (e.g., seat belts and child restraints) and devices that work automatically (e.g., seat-belt movement limiters, belt tensioners, and airbags). Modern seat-belt systems are extremely powerful and provide a high degree of passenger protection.

A crucial criterion for the quality of a restraint system is the perfect combination of the vehicle structure, the steering wheel movement, the seat performance, the occupant compartment layout, and the seat-belt characteristics.

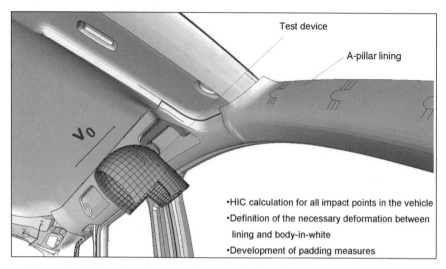

Figure 10.1 Finite element situation for head impacts, in accordance with FMVSS 201. (Source: Ref. 10-1.)

Modern computer simulation procedures allow the design of optimal occupant protection systems in cars.

10.2.1 Seat Belts

The standard design for the seating position of the front and rear occupant positions is the three-point automatic belt. The inner seat-belt latch is mounted for movable seats at the seat frame. The upper anchorage point is mounted at the B- and C-pillars. For several years in the United States, passive seat belts were standard equipment for the front seats.

The permissible field of the seat-belt anchorage points is defined by the legal specifications FMVSS 210 and the EC (European Community) directive [10-2, 10-3]. In nearly every car, the upper outside anchorage point is adjustable in the vertical direction for comfort reasons. Figure 10.2 shows the design of such a vertical adjustment.

A feeling of comfort goes with the protection function of the seat-belt system. This means that if the upper torso belt has a good geometrical position

*Figure 10.2 Seat-belt anchorage
height adjustment.*

in relation to the chest, the protection in an accident also is good. For the locking of the seat-belt retractor, two different mechanical systems are used. One system uses the acceleration or deceleration of the vehicle. Different kinds of pendulums are used as locking mechanisms. As a second type of locking device, the seat-belt pull acceleration is taken, where above a defined value (e.g., greater than 1g), the locking system must be achieved with a belt extraction of less than 50 mm (2 in.).

For the optimal combination of the seat belts with the vehicle structure, many subsystems exist.

The belt clamp locks the belt above the seat-belt retractor after the retractor is locked. With this feature, the remaining belt on the retractor is prevented from contracting more than 10 cm (4 in.), which is the length that could be exceeded due to the relative movement of the occupant to the vehicle. Figure 10.3 shows an example of such a design.

After the vehicle reaches or exceeds a certain deceleration level, the mechanical belt tensioner is activated by a spring, which pretensioned the belt in a time of 10 ms with a force of up to 2000 N. The mechanical belt tensioner, as shown in Figure 10.4, often is part of the belt latch.

Figure 10.3 Seat-belt clamp mechanism.
(Source: Autoliv.)

Today, more and more pyrotechnic belt tensioners are being used. These also have replaced designs such as the Audi PROCON-TEN system [10-4]. Depending on the severity of the accident, the pyrotechnic belt tensioner tightens the belt after exceeding a certain deceleration level. Figure 10.5 shows a tensioner that is mounted above the seat-belt retractor in the B- or C-pillar.

Other designs have the locking and tightening parts incorporated directly into the seat-belt retractor. In particular, the pyrotechnic seat-belt retractor allows greater freedom in the geometric design and in the layout of the motion/time function of the occupants.

New designs have incorporated an adaptive belt-force limiter with a dual-stage torsion bar [10-5], as shown in Figure 10.6. With this new design, the upper torso force level is reduced if the high force is acting for too much time.

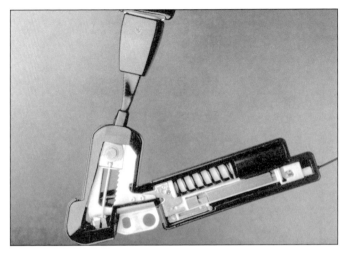

Figure 10.4   Mechanical seat-belt tensioner.
(Source: Autoliv.)

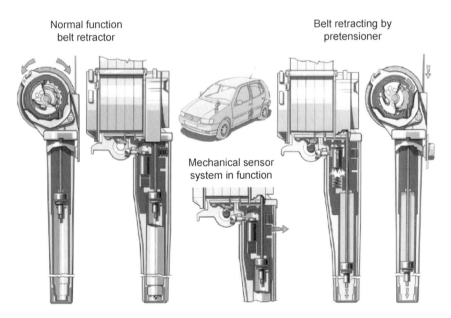

Figure 10.5   Pyrotechnic seat-belt tensioner.

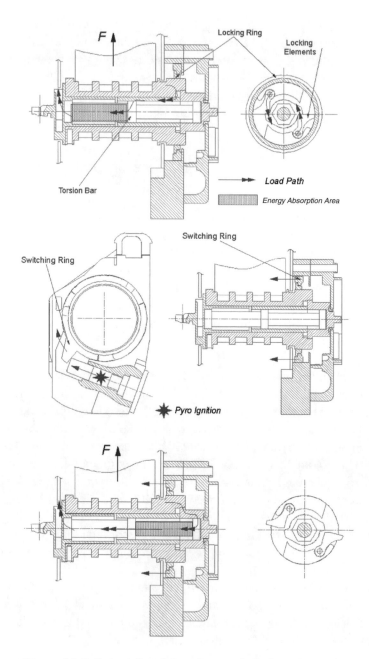

*Figure 10.6 Principle of the adaptive belt force limitation
with dual-stage torsion bar. (Source: Autoliv.)*

For occupants to be willing to use the seat belts, comfort is of great importance. The installation of the seat-belt latch at the seat has already made a positive contribution in that regard. Another design criteria is an optimization between the belt retraction force and the force to the occupant's chest in the worn condition. The minimum retraction force is necessary to guarantee perfect retraction. The retraction is influenced strongly by the belt material, the design of the upper belt buckle, and the belt guide above the retractor. Humidity also influences retraction, and some advanced designs use roller freewheels at the upper buckle to reduce belt/buckle friction. This consequently allows the reduction of the belt retraction force as well. Figure 10.7 shows an optimized design.

*Figure 10.7 Example of a run-through buckle
at the upper anchorage point.*

Other solutions for reducing the retraction force are on the market. For example, a locking mechanism always locks the retraction mechanism in the last position that the occupant has chosen.

With respect to comfort and seat-belt usage, passive seat belts were used in the U.S. market for more that 15 years. The reason for this development was the requirement by the National Highway Traffic Safety Administration for passive restraint systems. The rate of seat-belt usage was low in North

America, and the penalty for not wearing a seat belt was not introduced at that time. Furthermore, because airbag systems were not advanced technically enough to be sold, passive seat-belt systems were developed. These consisted of either passive two-point body belts with a knee bolster or passive three-point belts. The first car on the market with the passive seat-belt system was the Volkswagen Golf (see Figure 10.8).

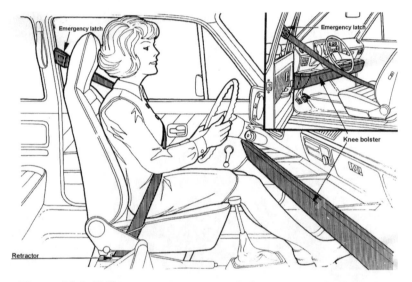

Figure 10.8 Volkswagen passive seat belt. (Source: Ref. 10-6.)

The body belt was attached on one side at the upper part of the door and on the other side at the seat, where the run-through buckle and retractor were mounted. Part of the belt also had an emergency latch and was attached via bolts to the seat-belt anchorage points at the B-pillar. If the latch was opened, the engine could not be started. To avoid "submarining" of the occupant under the seat belt, the seat plate was reinforced, and a knee bolster was installed in front of the occupants. In later versions, there also were some designs where the upper part of the seat belt was moved at the roof. If an occupant opened the door of the vehicle, the upper part of the seat belt was pulled into a forward position to allow easy entry and exit from the vehicle (see Figure 10.9).

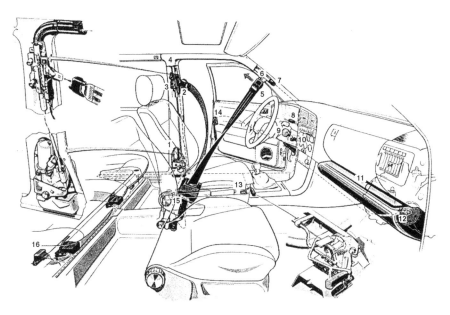

Figure 10.9 Layout of an automated passive seat-belt system. (1) Drive
unit; (2) Switch for B-pillar with latch control; (3) Locking device with
emergency latch and deformation element; (4) Transport rail; (5) Belt
latch; (6) Slip device; (7) Switch for A-pillar; (8) Warning light;
(9) Ignition; (10) Warning relay; (11) Knee bolster; (12) Switch rearward
gear; (13) Switch for sensor locking; (14) Door contact switch;
(15) Retractor with sensor locking; (16) Acceleration for driver and
passenger system. (Source: Ref. 10-6.)

With respect to avoiding submarining, many alternatives were available on
the market. These ranged from knee bolsters alone to actively operated lap
belts and a lap belt integrated into the automated three-point belt. With the
lap belt, the head and neck loading often increased because of the downward
pulling force created by the vertical forces of the lap belt.

Influenced by the progress in airbag development and performance and
because more U.S. states have introduced seat-belt usage laws, the passive
seat-belt system has been replaced by a combination of three-point belts with
airbag systems.

10.2.2 Airbags

Development of the airbag system began in the late 1960s [10-7]. At that time, the system used compressed gas as the deployment means. Although performance was good in accident simulation tests, the influence of outside temperatures changed the pressure of the airbag gas bottle significantly. At high temperatures, the pressure was too high; at low temperatures, the pressure was too low. In any case, airbags did not perform well. Either the bag pressure was too high, putting the occupant at risk of injury, or the bag failed to inflate. It took some time before a pyrotechnic solution was ready for introduction into production cars to avoid these disadvantages.

American legislative requirements to test airbag-equipped cars with and without the use of the front seat belt led to a complicated development of the airbag system design. The decision by the National Highway Traffic Safety Administration to define, via FMVSS 208, that the test should be performed also with unbelted dummies was based on the low seat-belt usage rates at that time, particularly in some U.S. states where no seat-belt usage laws were enforced.

We believe that airbags are only a supplement of the main restraint system: the worn seat belt. The public should not feel safe in any case that the vehicle itself gives adequate protection if occupants do not wear seat belts. Meanwhile, as we will see later, many additional requirements by FMVSS 208 have been introduced as of today, especially to cover the problem of occupants of different sizes, seating positions, and restraint system variations (e.g., when using child restraints). At the beginning of the rule-making period, airbags were considered for only frontal impacts.

10.2.2.1 Airbags for Frontal Impacts

If we analyze the performance of the three-point belt alone in frontal collisions, the occupant first moves forward relative to the vehicle interior compartment until the belt is locked. After that, the relative movement in relation to the body continues at a lower speed. Superimposed over this is a downward movement of the occupant due to the effect of the lap belt. This effect and the limitation of the belt elongation create at the final phase of the accident a stronger head rotation, which is reduced significantly by front airbags. Also, direct contact of the head against the vehicle interior is prevented. Figure 10.10 shows the layout of an airbag system for front occupants in frontal accidents.

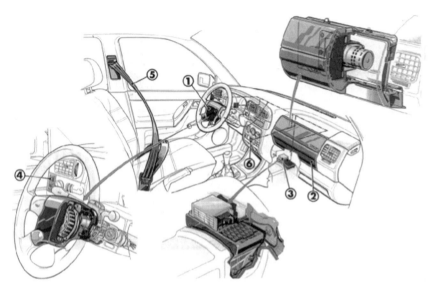

Figure 10.10 Layout of an airbag system. (1) Driver module; (2) Passenger module; (3) Control box with sensor; (4) Failure indication lamp; (5) Safety belt; (6) Diagnosis connection.

The system consists of several components: two pyrotechnic powder containers, the driver and passenger air bags with relevant covers, the electrical wire harness with several sensor units, the special power transmitter for the ignition of the steering wheel airbag, and the electronic control unit (ECU), including the diagnosis system. For gas generation, purely pyrotechnic powder or, in the meantime, some hybrid designs were used. The pyrotechnic material should create the least amount of emission components as possible, although the number of critical parts is small. The bag material consists of polyamid (PA) or polyester (PETP) partially with blow-out holes for a controlled ride downward. After a crash of a severity at slightly below the 20-km/h (12.5-mph) barrier speed, the ECU calculates in a threshold and deceleration time analysis whether or not the ignition of the inflation should be fired. After ignition, the pyrotechnic material creates a gas that fills the airbag. In a 50-km/h (31-mph) barrier impact, the sensor ignition time is less than 30 ms. The initiated airbag pressure lies between 1.8 and 2.2 bar. Figure 10.11 shows the principle function of an airbag system in connection with a three-point belt.

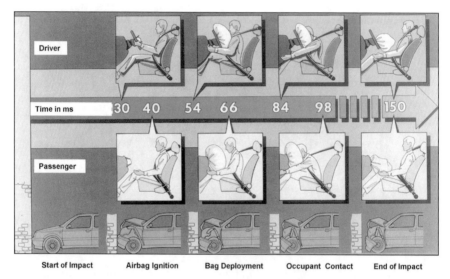

Figure 10.11 Inflation of driver and passenger airbags. (Source: Ref. 10-8.)

Because of the American experience with an airbag system that also must work both with and without a seat belt, the airbag pressure had to be much higher in the case where no seat belt was used. If the seat belt is used, basically only the head must be prevented from rotation. Without the seat belt, the mass that must be absorbed is more than 5 times higher. An easy comparison demonstrates this: the reduced head mass of a 50% male is approximately 6.8 kg (15 lb), whereas the reduced body mass is 36 kg (79 lb). This high pressure and the inflation speed of the airbags of up to 100 km/h (62 mph) has created several accidents with severe injuries. Figure 10.12 shows a distribution of lives saved versus deaths related to airbag performance, as of 2000.

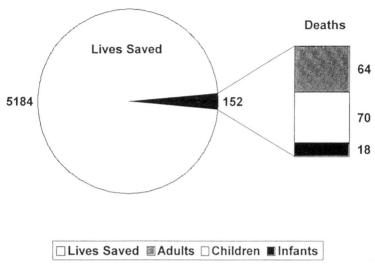

□Lives Saved ▦Adults □Children ▪Infants

Figure 10.12 Airbag safety effectiveness.
(Source: Ref. 10-9.)

New statistics show that in the meantime, because of improvements to the vehicle/airbag system, the number of deaths caused by inflating airbags has been reduced from a peak number of 56 in 1997 to 8 in 2001 (see Figure 10-13) [10-10].

In most cases, the severe injuries were consequences of incorrect use of the car seating position and seat belt. These can be categorized in the following groups:

• An out-of-position situation
• Wrong type of child seats and incorrect mounting position
• Children directly in front of the passenger front airbag

The immediate reaction to these accidents by vehicle manufacturers in most European countries was to allow the front passenger airbag to be disconnected (e.g., if child seats are installed). Meanwhile, some manufacturers offer dual-stage airbag systems. This means that in crashes with low severity, the bag inflation time is longer and the pressure is smaller than with high-speed accidents. Also, the standard FMVSS 208 has been revised to

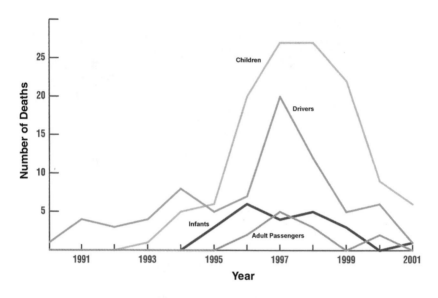

Figure 10.13 Deaths caused by inflating airbags, 1990–2001.
(Source: Ref. 10-10.)

a great extent, and European legislation and NCAP tests strongly influenced airbag design for frontal impacts. As a result of these changed requirements, we find that the airbag systems will have an even more complex design that is known as smart restraints. For frontal impacts, other types of airbags are under investigation, such as the knee bag. Revised FMVSS 208 of May 2000 [10-11] has many new requirements, which are shown in Figure 10.14 [10-12, 10-13].

The expansion of FMVSS 208 should increase the level of protection for a wider group of vehicle occupants and simultaneously minimize the risk for occupants of different sizes, occupants who are belted or unbelted, out-of-position passengers, and small adults, children, and infants.

If we compare these requirements with the rules of today due to the additional tests (e.g., with the 5% female Hybrid III dummy) and the necessity to prove the belted and unbelted cases, the experimental work also will increase. Furthermore, the additional definition and use of a neck injury criteria has increased the development effort.

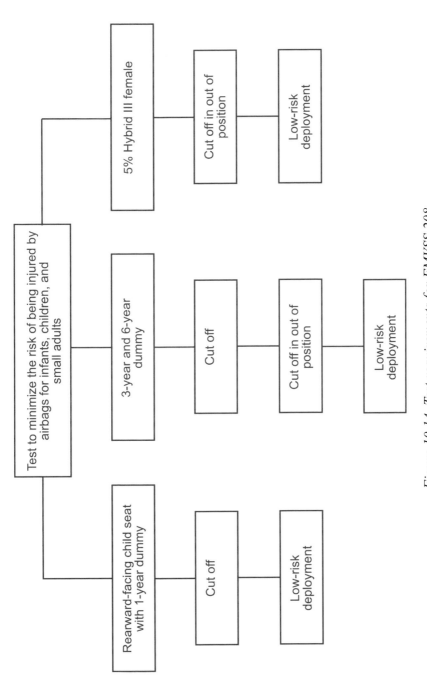

Figure 10.14 Test requirements for FMVSS 208.

For the second group (i.e., small adults, children, and infants), the requirements were defined as shown in Figure 10.15.

As stated, these new requirements should mean that a wider group of vehicle occupants will be protected. A specialty is the choice of the protection strategy. For the one-year-old child dummy, we can select the airbag cutoff or the low-risk deployment. In this case, the airbag is ignited with a performance stage that is determined in a pre-crash:

- For the one-year-old child dummy with a rearward-facing child seat with a 64-km/h (40-mph) test

- For the three- to six-year-old child dummies at the passenger seat and for the 5% female dummy with the 26-km/h (16-mph) crash against the fixed barrier.

Other options are cutoff in out-of-position situations and cutoff if the three- to six-year-old children are in front of the dashboard. This definitely means that the vehicle must be equipped with dual-stage airbags. Table 10.1 provides a good overview related to the complex test configuration and possibilities for complying with FMVSS 208.

Figure 10.16 shows one possible trigger strategy for the ignition of the airbag system, where the center and front sensors are used [10-13]. This strategy is related to the deployment threshold, where the obstacle, the closing speed, the impact direction, and the impact configuration play important roles. The legal requirements and the real-world accident situation strongly influence the deployment of air bags and the ignition of the pretensioner of the seat-belt system.

Although not requested by FMVSS 208, car manufacturers must test other occupant sizes, such as the 95% male. It is necessary to test not only the dual-stage design but the optimization of the airbag itself, size and ventilation holes, and the identification of occupants in the front seat. This means that the child seat must be recognized in the passenger seat, where the seating position of the passenger also must be identified, and several other techniques such as infrared radiation or ultrasonic must be used in combination with the seat sensor. Figure 10.17 shows a layout in principle.

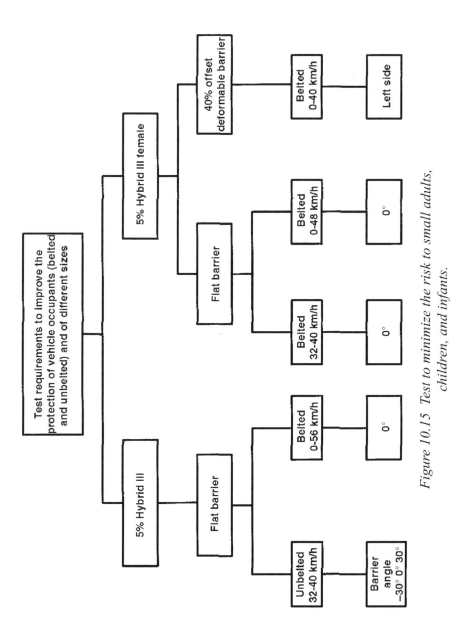

Figure 10.15 Test to minimize the risk to small adults, children, and infants.

TABLE 10.1
REQUIREMENTS OF FMVSS 208
AS OF SEPTEMBER 22003

Dummy/Test Requirement	50th Percentile Adult Male	5th Percentile Adult Female	6-Year-Old Child Dummy	3-Year-Old Child Dummy	12-Month-Old CRABI Dummy
Rigid Barrier Crash Test, Belted, 48 km/h (30 mph); Perpendicular	X*	X	N/A	N/A	N/A
Rigid Barrier Crash Test; Belted; 56 km/h (35 mph); Perpendicular	X	N/A	N/A	N/A	N/A
Rigid Barrier Crash Test; Unbelted; 32-40 km/h (20-25 mph), Perpendicular and 30° Offset	X	X (Perpendicular Only)	N/A	N/A	N/A
Offset Deformable Barrier Crash Test (Driver's Side); Belted; 40 km/h (25 mph)	N/A	X	N/A	N/A	N/A
Automatic Suppression Option (Static testing to determine if the airbag is automatically deactivated when an infant is in a car seat or a child is in the passenger seat in the position(s) specified in the standard)	N/A	N/A	X	X	X
Low-Risk Deployment Option (Vehicle must meet injury criteria specifications when the driver or passenger side airbag is deployed as specified in the standard)	N/A	X	X	X	X
Out-of-Position Dynamic Automatic Suppression Option**	N/A	X	X	X	N/A

*   48-km/h (30-mph) test using male dummies used for vehicles built through the production year beginning 9/1/06; after that production year, the high-speed rigid (56-km/h [35-mph]) barrier test using male dummies is required.

** Specifications for testing of the Dynamic Automatic Suppression Systems (DASS) have not been finalized by NHTSA.

Type A	Type B	Type C	Type D
Left and right UFS measure the same acceleration	Left and right UFS measure significantly different acceleration	Left and right UFS measure high acceleration, CCS measures low acceleration (- first 10-15 ms)	CCS, left and right UFS measure the same acceleration
• 100% barrier • Car to Car • Center pole	• Offset crash • Angular crash • Offset pole	• Truck under-ride	• Curb test • AZT

*Figure 10.16 Crash-type classification CCS end UFS sensors.
(Source: Ref. 10-13.)*

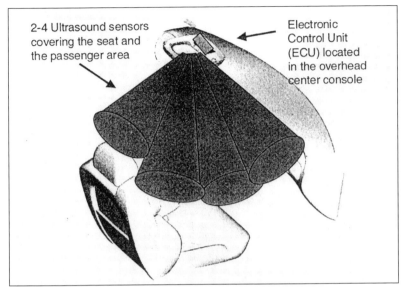

2-4 Ultrasound sensors
covering the seat and
the passenger area

Electronic
Control Unit
(ECU) located
in the overhead
center console

Figure 10.17   Ultrasonic sensors for occupant protection.
(Source: Ref. 10-13.)

10.2.2.2 Side Protection by Airbags

As a result of the great success of frontal airbags, side airbags in the upper torso area were introduced first by Volvo [10-14]. Special design features for this system are sensors for evaluation of the lateral acceleration and airbags, mostly installed laterally on the outside of the seat back. Because the free distance between the inner door and the occupants is much smaller in a side impact compared to frontal collisions, all events related to airbag performance must occur more rapidly. For example, the side impact sensor requires a strong connection between the lower part of the door and the outer part of the floor panel and the mounting location of the sensor. The results with the torso airbag alone were not efficient enough with respect to head protection in side impacts. For this reason, the head side impact airbag mounted at the outside of the inner roof was a positive step in further reducing injuries in side impacts. For side impact protection, because of the complex accident mechanism, a combination of seat belts, strong seats, good design of the occupant cell (i.e., integrity under high load), a stiff connection between the outer parts of the vehicle and the sensor mounting place, and the installation of a torso and head side airbag are the basis for good protection. Figure 10.18 demonstrates a head side airbag, which also protects the rear passenger, from the A-pillar as far as the D-pillar. Nearly the entire length of the windows is covered [10-15].

Figure 10.18 Side protection system in a pole impact crash simulation. (Source: Ref. 10-15.)

10.2.2.3 Additional Airbag Applications

In addition to front and side airbags, other airbag applications are possible, as follows:

• The side curtain airbag also could be used in rollover accidents to prevent head contact with the vehicle interior. Because a rollover takes more time (more than 5 sec) compared to milliseconds in frontal and side impacts, the airbag design must be changed in a direction to keep the airbag filled over a longer period of time.

• The knee area also has been investigated to reduce possible knee and femur injuries by the installation of a knee airbag at the driver and passenger seats. One initial application can be seen in the new BMW 7er (Fig 10.19) as of November 2001 [10-5].

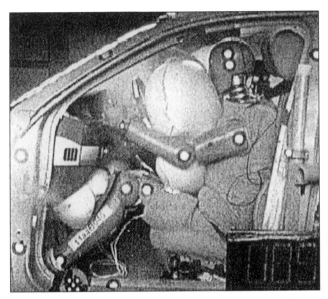

Figure 10.19 Knee airbags for the driver and front passenger. (Source: Ref. 10-5.)

• Another application has been investigated to fulfill future requirements for pedestrian protection. In this case, an airbag system could be installed at the front hood of the vehicle to reduce the deceleration level if the child and adult heads are impacting the front hood.

10.2.2.4 Sensors for Restraint Systems

Crash Severity. As shown in Figure 10.20, the number of sensors and actuators has increased continuously during recent years. In a Ph.D. thesis [10-16], it was found that the data from all sensors (front and side) could be used to better determine the crash severity and direction of the main impact

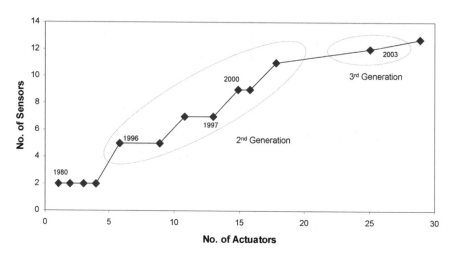

Figure 10.20 The number of sensors and actuators, as a function of years. (Source: Ref. 10-13.)

force. The rollover sensors provide an additional signal for this specific accident. In all relevant accidents, we must decide whether the airbag should be fired and at which level (Stage 1 or Stage 2), and whether the pretensioner of the seat belts should be activated (see Figure 10.21).

The vertical axis of Figure 10.21 shows the crash severity and the horizontal v-close. The v-close means the relative velocity between two colliding objects. In general, we could say that the combination of front and side sensors with a central sensor gives the highest degree of opportunities for correct determination of the accident severity.

For the reason already mentioned, pre-crash sensors also could be used in the future, such as a radar sensor that is working up to 10 m (33 ft) ahead and perhaps to the side. Some typical data are for the beams: 20° wide, 12° high, wavelength 905 nm, accuracy 0.1 m (0.33 ft), velocity range 4.8 to 190 km/h (3 to 118 mph) [10-17]. This might help to identify the type of closing obstacle. Because radar also cannot provide a determination with respect to mass and stiffness of the crash partner, the pre-crash sensor must be used in addition to the system already mentioned.

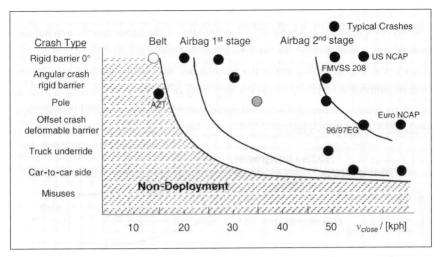

Figure 10.21 Typical deployment strategy for a dual-stage airbag. (Source: Ref. 10-13.).

Sensors in the Vehicle Compartment. The interior sensing systems must identify occupants (adults and children) on all seats. For the optimized performance of the airbag system, we need much more information about the vehicle occupants, such as why the airbag should be fired on the passenger side if no occupant has taken the seat, or, as another example, the passenger airbag should not be fired if a rearward-facing child seat is installed. Another important piece of information is the weight of the occupant, his or her size, and the seating position, particularly at the moment of the crash. Two types of sensor systems will be used in the future. One measures the pressure and the pressure distribution at the seat with a flat seat mat (see Figure 10.22), which can identify to some extent whether humans or child seats are occupying the relevant seat or whether the seat is empty.

The other type of sensor is used to determine the initial occupant position, because it is important to avoid situations in which the airbag could create negative effects. In this case, ultrasound or infrared sensors are the technical solutions.

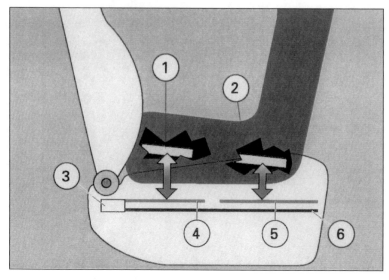

Figure 10.22 Child-seat recognition system on the passenger seat.
(1) Resonator #1; (2) Resonator #2; (3) Electronic device for child-seat
detection and passenger presence detection; (4) Antenna A; (5) Antenna B;
(6) Passenger presence detector. (Source: Ref. 11-4.)

Figure 10.23 shows a Photonic Mixing Device (PMD) that also can provide a determination about occupant movement [10-18]. The device has the following features:

- Emission of "RF" modulated incoherent light
- Reflection of desired three-dimensional object
- Detection of reflected light
- Mixing with originally emitted signal on chip
- Gray-scale image
- Distance information from each pixel

These sensors could suppress the airbag activation if occupants are in the non-firing zone. A typical situation is a child standing in front of the dashboard or if the occupant is out of position. Figure 10.24 shows a typical firing decision matrix for driver and passenger [10-13].

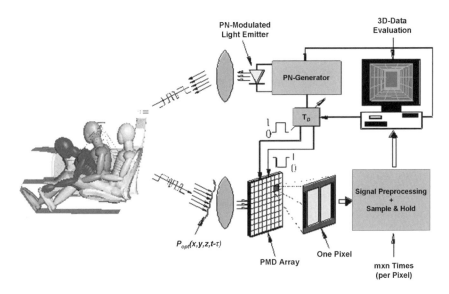

Figure 10.23 Photonic Mixing Device. (Source: Ref. 10-18.)

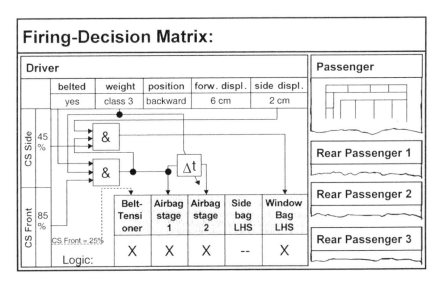

Figure 10.24 Firing-decision matrix. (Source: Ref. 10.13.)

Identification of Child Seats. In cases where the ISOFIX child seat is used, sensors in the two locking mechanisms also could identify a mounted ISOFIX child seat at the front passenger seat. In this case, the airbag in front of the child seat is not activated.

Figure 10.25 shows the large number of sensors and actuators for the different fields of application [10-19]:

- Accident avoidance
- Mitigation of injuries
- Driver assistance
- Autonomous driving

From Figure 10.25, we also can see that some sensor fusion seems possible in the future. It is important that driver assistant systems are developed and installed, which both support the driver and passenger and are accepted by them.

An advanced protection system was developed by DaimlerChrysler and is called Pre-Safe [10-20]. As shown in Figure 10.26, the system increases occupant safety after the system has recognized that a severe accident might occur.

Several protection items can be pre-activated. For example, the seat belt tensioner, the seat back is tilted into an upright position, the headrest is brought into its position, the sunroof could be closed, and vehicle parts such as armrests could be brought closer to the occupant. Although not all of these items might be seen in future production cars, the research approach is interesting. Some of these ideas are already in production.

10.3 Child Restraints

In accordance with legal requirements (e.g., in Germany), children who are younger than 12 years old and are smaller than 150 cm (59 in.) in height must be placed in special child restraints in vehicles. Child restraints must be designed in conformance with the regulation ECE-R44 [10-21]. Figure 10.27 shows an overview related to different child restraint systems as a function of the weight of the child [10-21]. The performance of the child restraint system is based on the limits defined in Table 10.2.

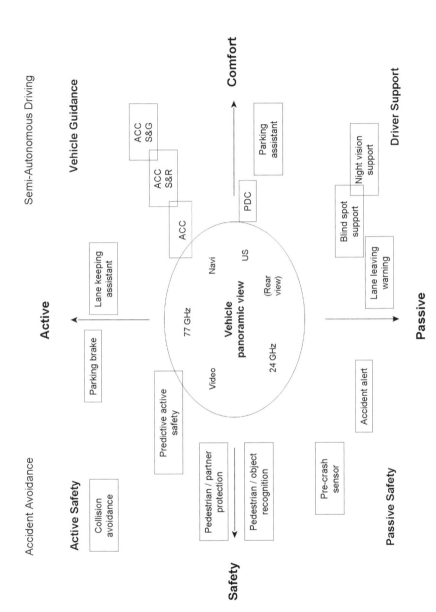

Figure 10.25 Sensor applications. (Source: Ref. 10-19.)

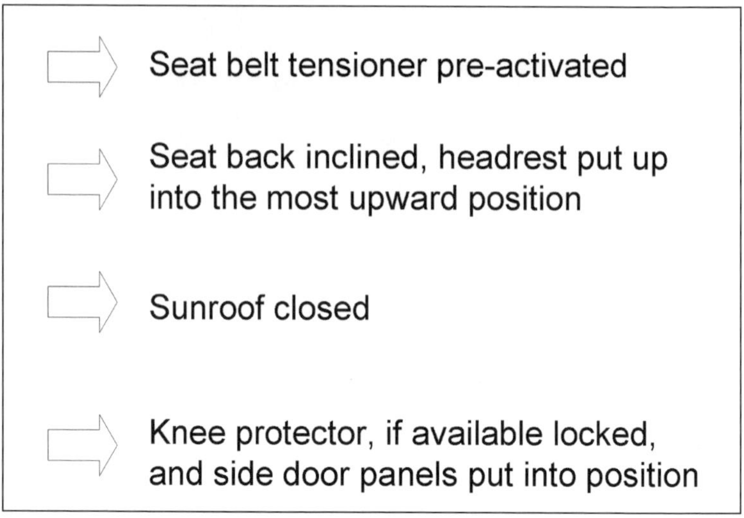

Seat belt tensioner pre-activated

Seat back inclined, headrest put up into the most upward position

Sunroof closed

Knee protector, if available locked, and side door panels put into position

Figure 10.26 Pre-Safe system.
(Source: Ref. 10-20.)

Special attention must be given to the correct use of child restraint systems. In competitive tests, it was shown, for example, that with a belt slack of approximately 75 mm (3 in.), the child dummy measurement data were approximately 30 to 40% higher, compared to a correctly used seat belt. Two special child restraints should be mentioned: (1) the baby carrier, and (2) the reboard seat. In most cases, both are placed on the front passenger seat. This automatically creates the previously mentioned conflict with the inflation of the front passenger airbag. The other system is the ISOFIX child seat [10-24], which includes in accordance with ISO [10-25] a standardized connection between the child seat and the vehicle interior. In particular, the two-point connection allows a good compromise between installation comfort of the child seat and the level of protection. In severe crashes, the two-point attachment also allows some rotation of the total seats. The big benefit of the ISOFIX is the reduction of installation failures of the child seat. Figure 10.28 shows an ISOFIX child seat from the Volkswagen group. It is very positive that in the meantime, automobile manufacturers are releasing their ISOFIX child seats for use with other car makers' models as well. For example, the BMW ISOFIX could be installed in Audi, Chrysler, Honda, Mazda, Mercedes, Peugeot, Renault, Toyota, Seat, Skoda, and Volkswagen models.

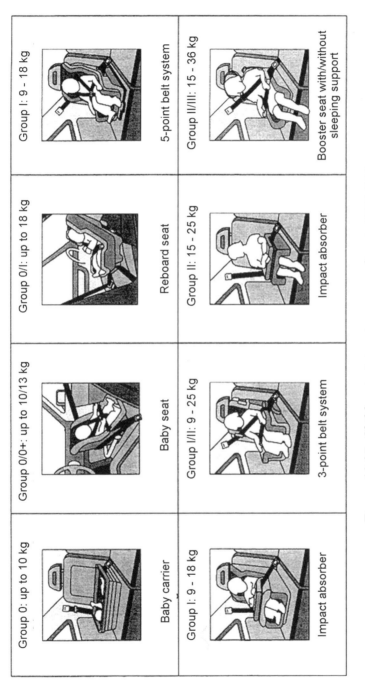

*Figure 10.27 Child restraint systems according to ECE-R44.
(Source: Ref. 10-21.)*

TABLE 10.2
REGULATIONS FOR CHILD SEATS

Limits for Children ECE-R44		Limits for Children CMVSS 213 [10-22], FMVSS 213 [10-23]	
Chest acceleration ($a_{res\ 3\ ms}$)	55 g	Head injury criteria ($HIC_{36\ ms}$)	1000
Chest acceleration vertical ($a_{z\ 3\ ms}$)	30 g	Head acceleration ($a_{res\ 3\ ms}$)	80 g
Head displacement horizontal	550 mm	Biomechanical values (experiences)	
		Neck moment (M_y)	20 Nm
Head displacement vertical	800 mm	Neck force (F_z)	2.0 kN

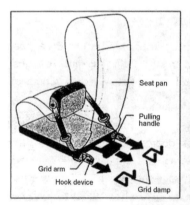
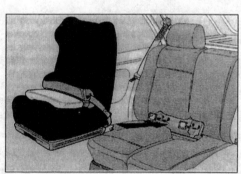

Figure 10.28 Schematic illustration of a two-point ISOFIX system. (Source: Ref. 10-23.)

178

10.4 References

10-1.   Oehlschlaeger, H., et al. "FEM-Crashsimulation: Ein modernes Werkzeug in der Nutzfahrzeug-Entwicklung," Proceedings of the Verband der Automobilindustrie (Association of the [German] Automobile Industry) (VDA) Technical Congress, IAA Internationale Automobil-Ausstellung [International Automobile Exhibition], Frankfurt, Germany, 2000.

10-2.   National Highway Traffic Safety Administration (NHTSA) and Federal Motor Vehicle Safety Standard (FMVSS), FMVSS 210, Seat belt assembly anchorages, National Highway Traffic Safety Administration, Washington, DC, United States.

10-3.   EC Directive, Brussels, 76/115/ECC "Seat belt anchorages."

10-4.   Vollmer, E. Audi AG, patent description EP 0189410 and EP 0191761.

10-5.   "Der Neue BMW 7er," *ATZ/MTZ*, Special Publication, November 2001, Verlag Vieweg, Wiesbaden.

10-6.   Seiffert, U. "Volkswagen Passive Occupant Protection Progress Report—1979," SAE Paper No. 790326, Society of Autmotive Engineers, Warrendale, PA, United States.

10-7.   Seiffert, U., et al. "Development Problems with Inflatable Restraints in Small Passenger Vehicles," SAE Paper No. 720409, Society of Automotive Engineers, Warrendale, PA, United States.

10-8.   Seiffert, U. *Fahrzeugsicherheit*, VDI-Verlag, Düsseldorf, 1992.

10-9.   Author's interpretation of data in information from the National Highway Traffic Safety Administration (NHTSA) and Insurance Institute of Highway Safety (IIHS).

10-10.   Status Report, Vol. 37, No. 4, Insurance Institute for Highway Safety, April 6, 2002, Arlington, VA, United States.

10-11. National Highway Traffic Safety Administration (NHTSA) and Federal Motor Vehicle Safety Standard (FMVSS), FMVSS 208, Occupant protection, May 2000, National Highway Traffic Safety Administration, Washington, DC, United States.

10-12. Szilagyi, I., et al. "Consequences of the new NHTSA regulations," Fifth International Symposium on Airbags, Fraunhofer-Institut, Karlsruhe, Germany, December 2000, ISSN 0722-4087.

10-13. Rölleke, M., et al. "Smart Sensors for Passenger Safety Systems," Fifth International Symposium on Airbags, Karlsruhe, Germany, December 2000, ISSN 0722-4087.

10-14. Volvo company, press release.

10-15. Scheef, I., et al. "Design of Side-Protection Features—Example Audi A4," Fifth International Symposium on Airbags, Fraunhofer-Institut, Karlsruhe, Germany, December 2000, ISSN 0722-4087.

10-16. Hübler, R. *Unterstützung bei der Auslegung von Airbagsystemen durch FEM-Berechnung*, Dissertation, Vol. 11, ed. by Institut für Elektrische Messtechnik und Grundlagen der Elektrotechnik, Technical University of Braunschweig, Germany, 2000, ISBN 3-89653-828-4.

10-17. Continental Temic Co. Product information material, 2003.

10-18. Schwarte, R., et al. "New Powerful Sensory Tool in Automotive Safety Systems Based on PMD Technology," Advanced Microsystems for Automotive Applications 2000, April 2000, Berlin, Germany.

10-19. Knoll, P. "Surround Sensing and Sensor Data Fusion," Verband der Automobilindustrie (Association of the [German] Automobile Industry) (VDA) Technical Congress, March 20–21, 2002, Stuttgart, Germany.

10-20. DaimlerChrysler, press release, IAA Frankfurt, 2001.

10-21. ECE-R44. "Einheitliche Bedingungen für die Genehmigung der Rückhalteeinrichtungen für Kinder in Kraftfahrzeugen," Economic Commission for Europe (ECE).

10-22. Canadian Motor Vehicle Safety Standard (CMVSS) 213.

10-23. National Highway Traffic Safety Administration (NHTSA) and Federal Motor Vehicle Safety Standard (FMVSS), FMVSS 213, Child restraint systems, Code of Federal Regulations, National Highway Traffic Safety Administration, Washington, DC, United States.

10-24. Langwieder, K., et al. "ISOFIX—Possibilities and Problems of a New Concept of Child Restraint Systems," in VDI-Berichte 1637, Vehicle Concepts for the 2nd Century of Automotive Technology, ed. by Verein Duetscher Ingenieure, VDI-Verlag, Düsseldorf, 2001.

10-25. ISO 13216-1 (1999). Road Vehicles—Anchorages in Vehicles and Attachments to Anchorages for Child Restraint Systems—Part 1: Seat Belt Anchorages and Attachments. ISO/TC 22/SC 12/WG 1 Child Restraint Systems (in Road Vehicles), ISO Secretariat, SMS-Sweden.

11.

Interrelationships Among Occupant, Restraint System, and Vehicle in Accidents

Although the single elements of the total vehicle, the vehicle interior, and the restraint system have been discussed in detail in Chapters 9 and 10, Chapter 11 describes the interrelationships among the occupant, the vehicle, and the restraint system.

11.1 Frontal Impacts

11.1.1 The Unrestrained Occupant

In a frontal impact against a fixed barrier or another vehicle, the occupant is moving toward the dashboard or the steering wheel due to the impact vector created by the impacting vehicle. In the first approximation, the following interrelationship is relevant.

Vehicle	Occupant Until Impact
$\ddot{s}_v = -a$	$\ddot{s}_{occ} = 0$

$\dot{s} = -a \cdot t + v_i$	$\dot{s}_{occ} = v_i$

Vehicle	Occupant Until Impact
$s_v = \dfrac{-a \cdot t^2}{2} + v_i t + s_o$	$s_{occ} = v_i \cdot t + s_o$

The relative movement Δs between the occupant and the vehicle can be calculated from the preceding formula as follows:

$$\Delta s = s_{occ} - s_v = \frac{a \cdot t^2}{2}$$

At a 50-km/h (31.15-mph) impact with an average vehicle deceleration of 15g and a distance between the occupant and the steering wheel of 30 cm (12 in.), the occupant hits the steering wheel after approximately 64 ms. Figure 11.1 shows the distance between the chest to the steering wheel in certain vehicles. From this figure, we could conclude that the 30 cm (12 in.) distance is not unrealistic.

The differential speed between the occupant and the steering wheel remains at 33.4 km/h (20.8 mph) at the time the chest hits the steering wheel because of the free forward movement of the occupant. In a more severe frontal accident without any restraint system, the kinetic energy of the occupant must be absorbed by the interior of the vehicle, (e.g., the steering wheel, the

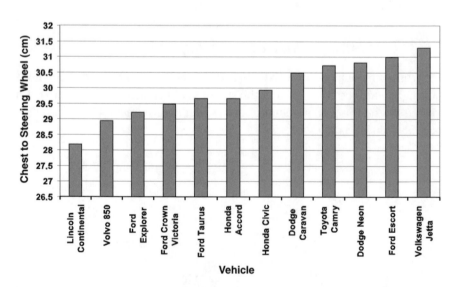

*Figure 11.1 Average chest-to-steering-wheel distance, by vehicle.
(Source: Ref. 11-1.)*

dashboard, the knee impact area, and the windshield). If we assume that for this deceleration, a deformation length of 0.1 m (0.33 ft) is available, then the average deceleration of the occupant is close to 50g via a time period of approximately 25.5 ms. Figure 11.2 demonstrates this theoretical pattern.

This deceleration level of the occupant body is already three times as high as the average vehicle deceleration level. This clearly explains two matters. If the occupant is hitting the vehicle interior, it is necessary to prevent injuries and, more importantly, to connect the vehicle occupant to a stable occupant cell as quickly as possible, to allow a ride downward in the restrained system together with the vehicle. For example, in the possible head impact zone, energy-absorbing material must be built into the vehicle interior. Figure 11.3 demonstrates one technical solution related to the new requirements of FMVSS 201. Of course, other energy absorbing materials, such as polyurethane foam, are possible.

The most important element with respect to minimizing injuries is the use of seat belts and/or the child restraint system.

11.1.2 The Three-Point Belt

With a three-point-automatic belt, without a belt clamping device or a pretensioner, the occupant is moving forward in the direction of the impact vector of the car that is being hit until the seat-belt retractor is locked either by the belt extension speed or by the force of gravity and until the belt is contracted at the belt retractor spool. Through the lap belt and the upper torso belt, the occupant's body is decelerated in its relative movement to the vehicle. With increasing load and the forward movement of the occupant as a result of its geometry, the lap belt creates a vertical load in a downward direction. For this reason, the seat and the seat pan must be integrated into the optimization of the performance of the overall occupant restraint system. Figure 11.4 shows, as a function of time, some typical data for the resulting head and chest acceleration and shoulder and pelvic belt force for a 50% male dummy that was restrained by a three-point-belt in a 50-km/h (31-mph) frontal barrier crash. As illustrated in Figure 11.4, it takes more than 30 ms until the belt is locked, any slack between the seat belt and the occupant is removed, and the first sign of an increase of the restraint force can be identified.

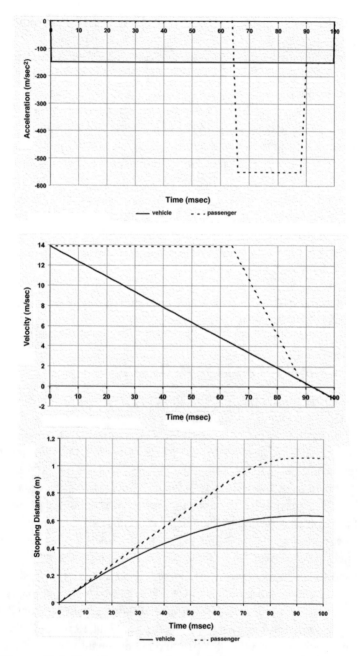

Figure 11.2 Deceleration, velocity, deformation, and movement, as a function of time.

Figure 11.3 Thermoplastic honeycomb for the head impact area.
(Source: Ref. 11-2.)

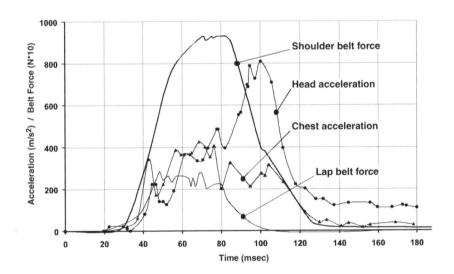

Figure 11.4 Decelerations and belt forces, as a function of time,
for a dummy restraint by a three-point belt.

For the performance of the restraint system, it is important to minimize this time as much as possible; therefore, we must prevent the pulling out of too much of the remaining belt from the retractor spool. The technical features used are belt clamping devices and pretensioners. In addition, the belt slack around the occupant must be reduced, especially when the occupant is wearing many pieces of clothing or heavy clothing (i.e., in winter). With the use of a pretensioner, this positive effect is achieved. After a specific accident severity level is reached, mechanical or pyrotechnic belt tensioners are activated. With the pretensioner, we also find some seat-belt load limiters to avoid the shoulder belt force being too high. Some of the systems used are described as follows:

11.1.2.1 The Seat-Belt Clamping Device Above the Retractor

This part can be used to stop a contraction of the remaining seat belt at the belt spool. By a kinematic activation through a relative movement of the occupant to the vehicle, the belt is grabbed directly above the seat-belt retractor.

11.1.2.2 The Mechanical Pretensioner

Basically, two systems were put into production: (1) a pretensioned spring with a mechanical or a pyrotechnic release, and (2) as a specialty, a system called PROCON-TEN, which was used by Audi AG for several years and is shown in Figure 11.5 [11-3]. This system used the relative movement of the transmission housing and the vehicle body to pull the steering wheel away from the driver, and tensioned the belts for the driver and right front passenger, if the accident was severe. The belt forces in the shoulder belts were limited by force limiters to avoid chest injuries.

Most of the devices mentioned here are no longer in use today because the pyrotechnic pretensioner is more flexible in its vehicle applications and better in its performance.

11.1.2.3 The Pyrotechnic Pretensioner

Again, several different designs are in production. Most often, a pyrotechnic pretensioner is used either as part of the seat belt, above the seat-belt retractor, or integrated into the retractor itself. Figure 10.5 shows a typical layout for a pyrotechnic pretensioner.

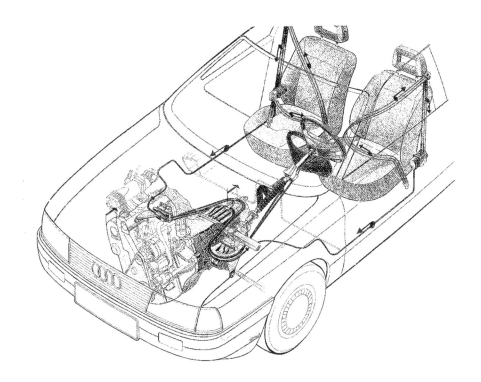

Figure 11.5 The PROCON-TEN system by Audi AG. (Source: Ref. 11-3.)

The pretensioning function is activated by a sensor-triggering signal. This sensor could be a mechanical or an electronic sensor, together with the relevant electronic control unit. With this technical feature, the designer has more capability to improve the performance of the restraint system.

Figure 11.6 shows another design [11-4]. This design of the pretensioning device has been integrated into the seat-belt retractor and has a function similar to the rotor of the Wankel engine. In a severe crash, the spool of the retractor is rotated via a pyrotechnic gas in the direction opposite to the pull-out function, if the sealt belt is worn by the occupant.

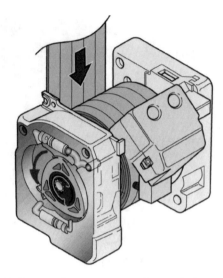

Figure 11.6 Pyrotechnic rotational seat-belt pretensioner.
(Source: Ref. 11-4.)

11.1.2.4 The Seat-Belt Load Limiter

To prevent too high a seat belt load, especially at the shoulder belt, various three-point belts use an integrated seat-belt force limiter. Figure 11.7 shows a relatively simple layout [11-4].

By the selecting the proper seat-belt retractor dimensions and the rotating shaft material, the force level can be predetermined. The most sophisticated system, as of today, is installed in the new BMW 7 series. The principle for this system has already been shown in Figure 10.6. Figure 11.8 shows the load level as a function of time [11-5].

The adaptive belt force limiter uses a two-stage torsion bar installed via the shoulder belt part in the seat-belt retractor. With a pyrotechnic device, the system can switch from a high force level to a low force level. In the initial stage, the high force level is applied. If the accident is severe, the system switches during the loading phase, based on the force applied by the occupant to the seat belt. This allows a force time function where the force is reduced, if the loading phase that is applied through the occupant to the upper torso belt takes too much time.

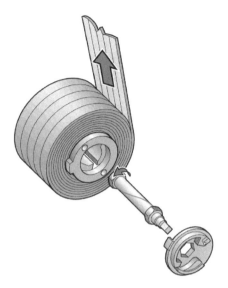

Figure 11.7 Seat-belt force limiter.

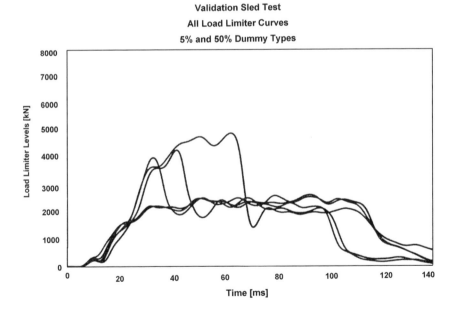

*Figure 11.8 Adaptive belt force limitation with a two-stage torsion bar.
(Source: Ref. 11-5.)*

11.1.3 Passive Restraints

Because of the low seat-belt usage rate, especially in the United States in the late 1960s, the introduction of passive restraints into production cars was discussed. One solution at that time was the installation of airbags for the front passenger. At that time, the filling process for the airbag was done by compressed gas. Unfortunately, the pressure in the gas bottle was too sensitive to temperature changes. This means that in the case of high temperatures, the bag pressure was too high. Likewise, at cold temperatures, the bag did not inflate sufficiently. Even in later years, especially in the United States because there were no special campaigns or enforcement actions, the seat-belt usage rate was not high enough. As a consequence, the request for the introduction of passive restraints increased again.

Figure 11.9 shows for some individual U.S. states an increase of approximately 10% by buckle-up campaigns, with a usage level between 49 and 84%. Because of the problems with airbags as mentioned, passive seat belts

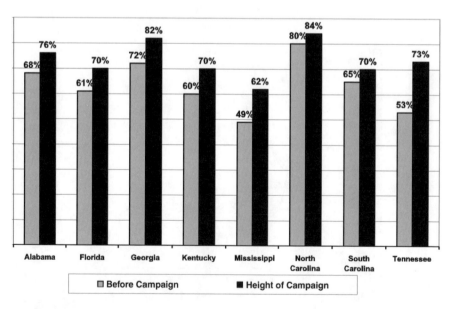

Figure 11.9 Seat-belt usage rates in various states in the United States.
(Source: Ref. 11-6.)

were installed in various vehicles in the United States and partially in Europe as an interim solution.

The Volkswagen Passive Restraint System [11-6, 11-7]. The Volkswagen Restraint Automatic System (VW-RA) was the first such system in the U.S. market. Passive seat-belt systems were either upper torso belts with a knee bolster and/or, at a later stage, an additional lap belt or passive automatic three-point-belts. Figure 10.8 shows the design of the VW-RA.

The upper torso belt was connected on one side via an emergency latch to the door frame and on the other side at the seat structure. In the event of an accident, two bolts transmitted the belt pulling forces to a plate that was mounted at the B-pillar, where normally the upper seat belt anchorage point is installed. In front of the front occupants, a special knee bolster prevented the "submarining" of the occupant during frontal accidents. The anti-submarining effect also was supported by the seat layout, where particularly the seat pan was designed strong enough to prevent submarining. Figure 11.10 shows a comparison of the head and pelvic motions for a three-point-belt compared to the VW-RA.

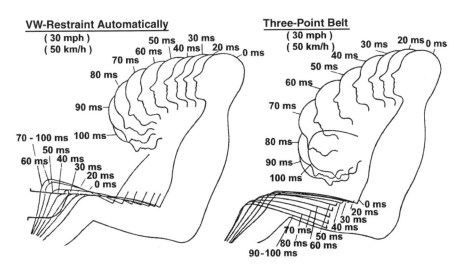

Figure 11.10 Difference of occupant kinematics for a passive body belt and conventional three-point belt. (Source: Ref. 1-1.)

As shown in Figure 11.10, we could observe significantly less head rotation and vertical ride downward for the occupant restrained by the VW-RA. The VW-RA was offered in a subcompact car. For compact and high-class cars, passive automatic three-point belts were installed. If a lap belt was used, the performance was more similar to the behavior of a three-point belt.

The upper torso belt did move automatically into a forward position if the front doors of the vehicle were opened, and it returned to position after the doors were closed. All other elements such as the pretensioner and force limiter, as well as an emergency locking mechanism, were possible.

Although the degree of comfort was much higher than with the standard three-point belt, the further development of the airbag system caused a change in restraint system technology.

11.1.4   The Airbag System

Since first introduced in production vehicles during the second half of the 1980s, the airbag system has undergone continuous development that is ongoing today. Figure 10.10 shows the layout for a basic airbag system.

The sensor system identifies the crash severity and, for some vehicles, the occupant seating position, as explained in Chapter 10. For the driver, the airbag is installed in the center of the steering wheel. For the passenger, the airbag is installed in the upper part of the dashboard. Because of U.S. requirements for the fulfillment of FMVSS 208, the airbag system also must be tested in the relevant vehicle without seat belts, and the lower part of the dashboard is designed in such a way that in the event of a leg impact, no sharp edges nor rigid vehicle parts create injuries. Furthermore, some energy absorption to avoid too high of an upper load must be possible. After the evaluation of the sensor signal through the electronic control unit, and if the decision is made that the airbag should be deployed, the pyrotechnic material in the gas generator is ignited, and the gas fills the driver and front passenger airbags. The size of the driver airbag is approximately 60 l (16 gal), and the passenger bag is approximately 120 l (32 gal). In most cases, the bag material is normally polyamid with some pop-out holes to allow a better ride downward of the occupant. The geometric layout depends greatly on the vehicle itself. On the driver side, we frequently find a cylinder shape for the airbag. On the passenger side, the airbag resembles the form of a tube. Inside the airbag in some designs are strips to prevent the airbag from becoming an undefined shape. Especially on the passenger side, the airbag configuration also must take into consideration the performance in an accident with an out-of-position occupant.

In a single-stage system, the bag can reach a relative high speed of deployment of more than 100 km/h (62 mph), when it comes close to the occupant. In a dual-stage mode, the airbag is filled with lower pressure if the accident is not as severe or if the occupant is located in an out-of-position situation. One example is a standing child or if the occupant is too close to the dashboard. Also, different occupant sizes, such as a small driver sitting too close to the steering wheel, influence the strategy of the airbag inflation characteristics and the geometric design.

With respect to the pyrotechnic material, improvements were made to reduce particles and carbon monoxide. New designs also are using a hybrid generator as a good compromise between the inert gas and the propellant material, as shown in Figure 11.11.

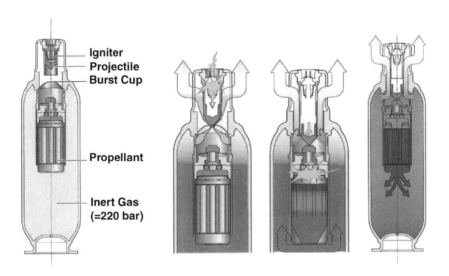

Figure 11.11 Hybrid gas generator. (Source: Ref. 11-4.)

To reduce the knee and upper leg forces and the local pressure at the knee, the first production cars also featured a knee bag installed below the steering wheel area.

As mentioned, the airbag system is only complementary to the major restraint system, which is the seat belt. The optimal protection function can be achieved only via a good combination of the crash characteristics of the vehicle structure, the integrity of the passenger compartment, the seat, the seat-belt system, and the airbag.

11.1.5 Comparison of Test Results for Different Restraint Systems

In Figure 11.12, dummy measurement data of two different restraint systems in a frontal collision are compared. In one case, we can see a three-point belt; the other case shows a combination of a three-point belt with a driver airbag.

11.2 Lateral Collisions

The possibilities for applying special restraint systems for the protection of vehicle occupants in lateral collisions are much smaller compared to those for frontal collisions. This is due to the proximity of the occupant to the vehicle interior side (door). One of the influencing parameters is the stiffness of the body-in-white. The side of the impacted vehicle must be strong to show at least some resistance against the intrusion of an impacting car. This requires a corresponding design of the body sills, the side beams in the door, and other door reinforcements (e.g., a connection via a hook between the lower part of the door and the body-in-white sill). Also, cross-bars (i.e., below the dashboard) in the vehicle and the seat design in the seat-mounting area have a positive effect on crash protection with respect to resistance against lateral forces, as well as cross-bars in the rear of the vehicle. The other parameter is the layout of the side interior of the vehicle. Special attention must be given to the size, shape, and material characteristics of the side door panel and the head impact zone.

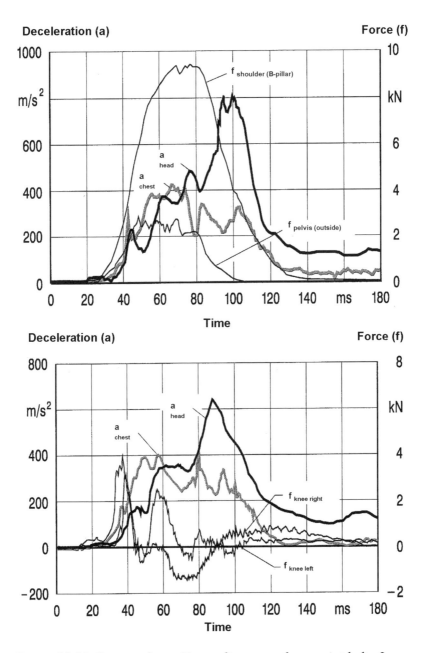

Figure 11.12 Dummy data. Upper diagram: three-point belt. Lower diagram: three-point belt with airbag (driver).

The newest element of passenger protection in lateral collisions is the installation of side airbag systems in the chest and head impact areas.

11.2.1 Theoretical Analysis

Compared to frontal impacts, the variables of an accident pattern are much higher in lateral collisions, such as the direction of impact, point of impact, and geometry of the obstacle. The main parameters are as follows:

- Collision partner involved (mass, deformation forces, structural integrity, and structural geometry)

- Occupant size

- Impact point and angle

- Impact velocities

- Seating positions of the occupants

- Interior design of the vehicle

- Use of the seat-belt system by occupants

- Installation and type of side airbags

For a better understanding of this complex accident situation, Figure 11.13 explains in principle the sequence of a typical lateral collision between two vehicles. With a simplified representation of the impacted and impacting vehicles and the one simulated occupant, the velocity–time function of the two vehicles and the occupant is shown.

In Figure 11.13, the numbers 1 to 6 identify the velocity–time function related to the occupant and parts of the vehicle. The areas surrounded by these velocity traces correspond with the relative distance between the identified vehicle parts and between the vehicle components and occupants. The free deformation characteristics of the structure and the lateral restraint system—in this case, a cushion—have been defined to react against external forces

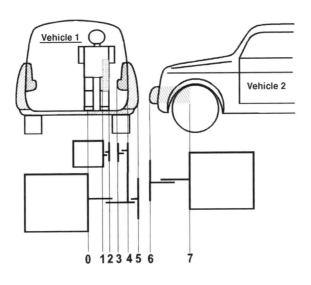

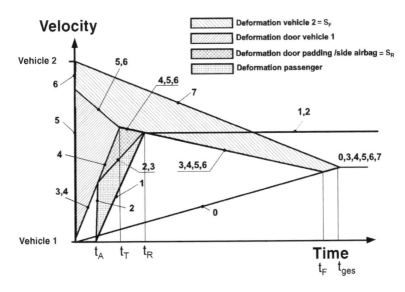

Figure 11.13 Theoretical velocity–time functions for the vehicle structure and the occupant in side impacts. (Source: Ref. 1-1.)

with a rectangular force deflection characteristic. Also, the oscillation of the dummies or vehicle parts is not taken into consideration in this simulation. Therefore, the results for the velocity v, the deformations, and the relative movement of the simulated occupant as a function of time f(t) differ from the behavior in a real accident. However, the main interaction of the occupant and the vehicle can be understood. The impacting vehicle drives with the velocity under 90° into the side of the impacted vehicle. In this case, the impacted vehicle has zero velocity. After a short time, the outside of door 5 reaches the same velocity as the bumper of the impacting vehicle 6. The inside of door 3 reaches the same velocity as the bumper. The contact between the occupant and restraint system 3 starts at the time when the occupant hits the inner door. At the time t_T, the relative velocity between occupant and door reaches its maximum. The relative speed could become higher as the total change of velocity of the impacted vehicle. The occupant is decelerated by the inside components of the vehicle. The deformation of the door, including the armrest, is s_R. The occupant and the bumper have at time t_R the same velocity. Until point t_{ges}, the velocity of the impacted occupant remains high until the occupant comes to a standstill or hits the front-seat passenger. The deformation of the side structure is finished at the point t_F because the seats and the underbody of the vehicle, including the cross-bars, have no further deformation capability. The front structure of the impacting vehicle has deformed itself by the amount of s_F. The deformation of the side structure is described through the relative movement of door 5 related to an undeformed point of the impacting vehicle 7. Figure 11.14 shows that even without airbags, a strong relationship is given by different layouts of the structure and the vehicle interior, especially the door.

The upper part of Figure 11.14 shows a design layout with a relatively low stiffness of the side structure combined with some thin upholstery of the impacted vehicle. This means that a relatively large free distance exists between the occupant and the upholstery. For the occupant, high loadings are expected if the distance between the occupant and the inside of the vehicle increases. The upholstery is deformed at point t_A. The occupant then is spontaneously accelerated to the speed of the front part 6 of the impacting vehicle. Because of the low energy-absorption capability of the upholstery, this impacting speed cannot be reduced.

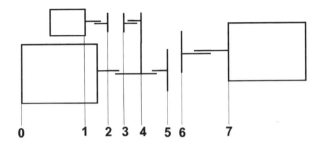

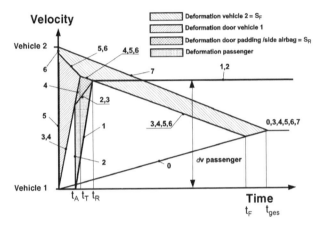

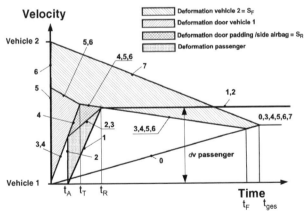

Figure 11.14 Theoretical velocity–time functions for the vehicle structure and occupant, with vehicle variations in side impacts.

The lower part of Figure 11.14 shows the other extreme. In this case, the side structure of the impacted vehicle is stiff, and the upholstery at the door is relatively large. It also could be a torso side airbag. Loadings for the occupant are much less because of the energy-absorbing capability of the upholstery or the torso side airbag and the lower relative velocity of the inner door. This is because the contact between the occupant and the inner part of the door occurs in a much shorter time.

11.2.2 Side Airbag Systems

11.2.2.1 Thorax Side Airbags

The first thorax side airbag was installed by the Swedish car manufacturer, Volvo [11-9]. Figure 11.15 provides a good overview of the system. This system consists of an airbag installed in the outer side of the seat back, a pyrotechnic inflator, and a sensor with the control unit.

Initial results in comparative testing showed in some cars similar deceleration levels at the chest area, with and without installed thorax airbags. This is easily understandable if we examine the distance between the inner door panel and the occupant. Even with a thorax airbag, the free deformation length in the transverse direction is rather small. On the other hand, if we compare the specific pressure at the occupant or at the dummy due to the much larger impact area typically offered by the thorax airbag, there is an advantage for the airbag system. This means the deceleration level is not reduced, but the specific pressure is much lower. Figure 11.16 shows the results of a computer simulation for the analysis of the protection level for a child in a lateral impact, with and without an airbag [11-9]. The improvement is significant, particularly with the optimized version. The two systems in production today are torso bags, mounted either in the outer side of the seat back or in the inner part of the door.

The thorax airbag also allows earlier contact by the impacting door panel. The acceleration distance for the impacted occupant will be extended through this effect. The seat-mounted version has the advantage, in that the position of the thorax airbag is independent from the seat longitudinal position. This provides not only a constant position to the occupant but allows more rapid

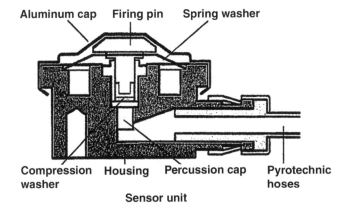

Aluminum cap   Firing pin   Spring washer

Compression   Housing   Percussion cap   Pyrotechnic
washer   hoses
Sensor unit

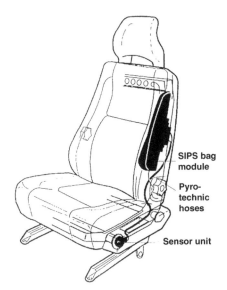

SIPS bag
module

Pyro-
technic
hoses

Sensor unit

Figure 11.15 Vehicle seat with SIPS bag module.
(Source: Ref. 11-4.)

inflation because of the smaller size of the airbag. Figure 11.17 shows the
time sequence for airbag performance in a lateral collision [11-10].

The side airbag, including the sensor control that calculates the level of
severity of the accident, is inflated after 15 ms. The occupant is hit after 20

Protection Potential
SINCAP, HIII 3y

—— Without airbag —— Basic —— Low Risk

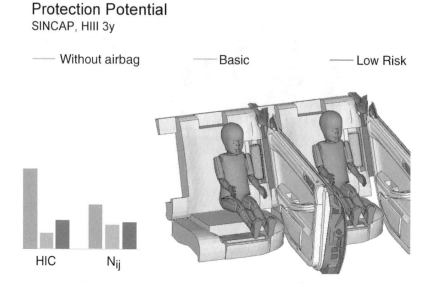

Figure 11.16 Comparison of the protection level for a child dummy in a lateral impact, with and without an airbag. (Source: Ref. 11-9.)

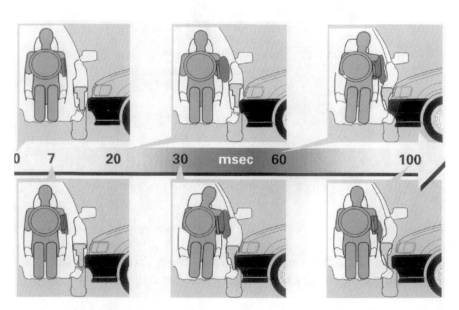

Figure 11.17 Typical side airbag function versus time. (Source: Ref. 11-10.)

to 25 ms. In a side impact, the occupant can be supported by the padding design in the pelvic area, as well as through the airbag in the chest area. Some other solutions are the use of a special shape of side airbag, the thorax-pelvic airbag. The decision regarding which type of airbag will be used depends on the strategy of the vehicle manufacturer. Figure 11.18 shows the two different installation modes [11-14].

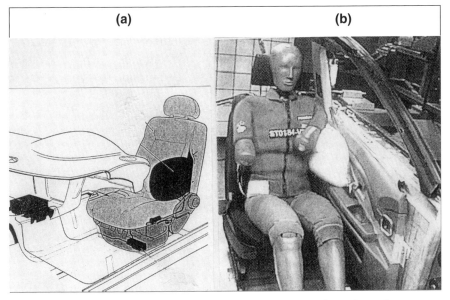

Figure 11.18 Two different modes of installation for side airbags:
(a) seat-mounted; (b) door-mounted.

For lateral collisions, special attention must be given to the sensor installation. The sensors must be as close as possible to the doors; however, they also must be located in the area where the passenger compartment remains intact, even in severe accidents. The free distance between the side door sill and the sensor should be as small and as stiff as possible to transmit the acceleration data created by the impacting car to the sensor unit as rapidly as possible. New sensors that measure pressure during contact with the impacting vehicle are under development.

11.2.2.2 Side Head Protection Airbags

After installation of the torso side airbag, solutions for head protection in lateral collisions became more important. Figure 11.19 shows an analysis of severe injuries for different regions of the human body in lateral collisions. The data were taken from cars before the head protection airbag was installed. These data show that injuries are most common to the head, thorax, and neck. Again, several installation modes for head airbags are possible, as shown in Figure 11.20 [11-4].

Meanwhile, the "inflatable curtain " has become an attractive installation system. The head airbag is installed in the side roof area as a pre-assembled unit to protect the heads of the outside front and outside rear passengers in severe lateral collisions. Figure 11.21 [11-12] demonstrates deployment of a typical window airbag, which prevents severe head impact against the vehicle

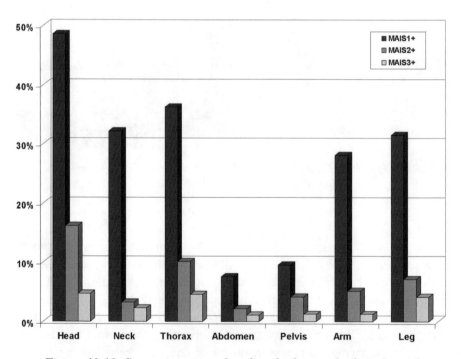

Figure 11.19 Severe injuries related to the human body in lateral collisions. (Source: Ref. 11-11.)

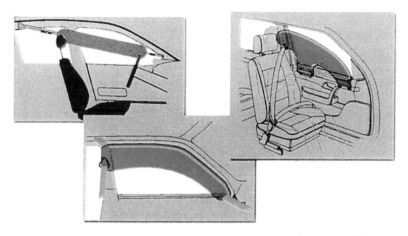

*Figure 11.20 Different head airbag systems for lateral collisions.
(Source: Ref. 11-4.)*

Figure 11.21 Side window airbag. (Source: Ref. 11-12.)

interior such as the B-pillar and the pop-out of the occupants' heads through the side window area.

11.3 Rear-End Collisions

For the development of vehicles related to their performance in rear-end accident simulation tests, in addition to the full-scale test (check of the integrity of the fuel tank and fuel lines), the resistance of the seats, and the performance of the seat backs are tested most frequently in sled tests and in tests that are defined by the vehicle manufacturer. The legal test procedures for the seat and head restraints are valid for the single components but not for total vehicle tests. Although the number of rear-end collisions is small compared to the number of frontal and lateral collisions, the whiplash injuries that occur in rear-end collisions are complicated and the injured people take a longer period to recover. Rear-end collisions also represent a high percentage of cost with respect to insurance claims. Therefore, the newest features for occupant protection in rear-end collisions also are based on the results of accident statistics. In a study conducted by Temming [11-13], it was found that females have a higher risk compared to male occupants and that a large number of injuries occur (see Figure 11.22) at a relatively low change in velocity of approximately 8–12 km/h (5–7.5 mph).

For evaluation of the performance of the seat and the seat head restraints, many mathematical models exist, and new types of three-dimensional dummies have been developed. For example, many activities conducted in Sweden [11-14 through 11-16] resulted in the development of the Bio-RID dummy. This dummy has a multi-segmented spine and allows better determination of the performance of the seat, including the head restraint. A paper by TNO Automotive [11-17] describes the RID 2-2 prototype, which was developed in the European whiplash project. The results definitely will improve the situation with respect to the evaluation and assessment of the vehicle, the seat, and the head restraint related to their performance in rear-end collisions.

The physical relationships among the partners in rear-end crashes are described briefly here. In addition to some plastic deformation, the impacting car is creating a change in velocity of the impacted car by

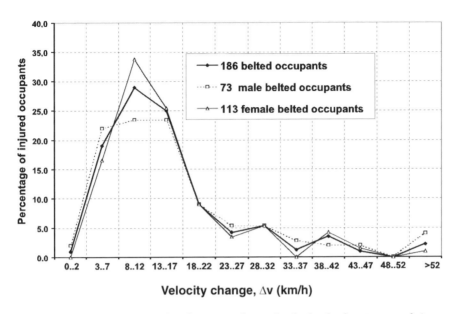

Figure 11.22 Relationship between the risk of whiplash injury and Δv (single rear-end impact, belted occupants). (Source: Ref. 11-13.)

$$\Delta v_c = v_r \cdot \frac{m_i}{m_c + m_i}$$

where

Δv_c = Change of velocity of the crashed car
v_r = Relative speed
m_i = Mass of the impacting car
m_c = Mass of the crashed car

The acceleration level of the impacted vehicle usually is significantly below the deceleration level observed in frontal collisions because the free deformation zone in the rear of the impacted vehicle is larger on average, and the relative velocities are smaller compared to a car-to-car frontal collision. Figure 9-13 shows that even in a collision with a speed of 48.3 km/h (30 mph) and a heavy barrier (1800 kg [3968 lb]), the maximum acceleration, measured at the impacted car, is in the range of only 15g, compared with a frontal impact with a fixed barrier of only 1/3 (see also Figure 9-4).

The forces created by the inertia force through the occupant are transmitted first to the seat back, then to the seat frame, and (because of the rearward bending of the spine) to the headrest. Some basic requirements must be fulfilled. It is evident that the seat back should not bend too much, so that the head contact is so late that the head does not receive any support from the head restraint. Another problem will occur if the headrest is too low. In this case, it might increase the injuries to the occupants. Data are published from some institutions (e.g., in Switzerland and also by IIHS [11-18]), which represent the results of head restraint with respect to geometric data. Figure 11.23 demonstrates the measurement methods and the methods of the judgment with respect to geometric data as performed by the Insurance Institute for Highway Safety (IIHS).

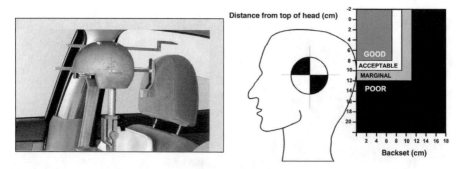

Figure 11.23 Head restraint measurement. (Source: Ref. 11-18.)

As mentioned, in rear-end collisions we experience a dynamic loading of the seat, the seat back, and the head restraint. Thus, the real performance is shown only in dynamic tests. Because whiplash-associated disorders are so significant in the field, car manufacturers have reacted on their own. One indication of that activity is shown in the geometric layout of the head restraints. Figure 11.24 demonstrates that the head restraints rated as "good"

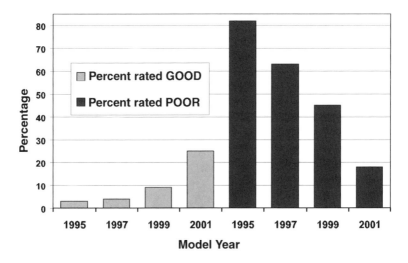

Figure 11.24 Head restraint improvements.
(Source: Ref. 11-18.)

by IIHS increased from 3.3 to 28.7% between 1995 and 2001, and the head restraints rated as "poor " decreased from 81.7 to 22%.

Another innovation was created by Saab [11-19]. By active loading of the seat back, the head restraint is brought into its most upward position. The Saab active head restraint system (SAHR-System) has been in production since 1997. Figure 11.25 shows its basic function.

Also, biomechanical activities such as the development of the Bio-RID dummy have contributed to improvements of the seats, including the head restraints— not only in laboratory tests but with respect to their performance in the field. If we look to the neck extension angle as a function Δv, there is significant improvement in sled tests and also on the roads. The analysis of accidents on Swedish roads shows a reduction of 75% of accidents with people who received whiplash injuries, after the SAHR-System was installed [11-20]. The study was performed two years after the SAHR-System was in production.

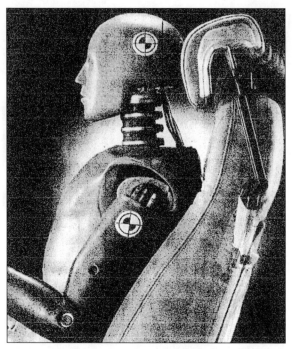

Figure 11.25 Headrest developed by Saab.
(Source: Ref. 11-19.)

11.4 Rollover Protection

The vehicle body with its designed structure, the vehicle interior, and also the glued-in front windshield provide good protection to vehicle occupants during rollover accidents, particularly if the occupants are wearing seat belts. For example, the glued-in windshield increases roof resistance during the quasi-static roof test (FMVSS 216 [11-21]) by more than a factor of two [11-22]. Also, it is obvious that the doors of the vehicle should not open by themselves during the rollover, and a valve activated by gravity forces in the ventilation line from the fuel tank to the charcoal canister prevents fuel leakage. Figure 11.26 shows head acceleration during a dynamic rollover test, described in FMVSS 208, for two 50% male dummies on the driver and passenger sides. As these data show, the resultant head acceleration is less than 20 to 25g, which is far below the critical value of 80g.

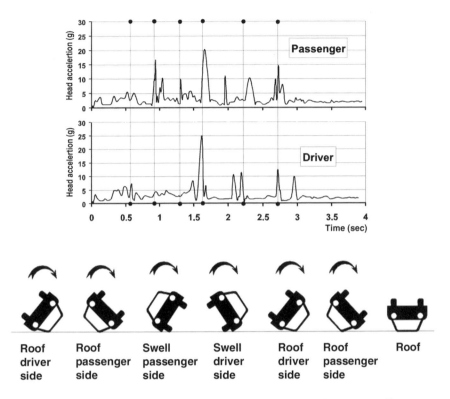

Figure 11.26 Dummy measurement data during a dynamic rollover.

For more severe accidents, the occupants' heads also must not come into contact with other obstacles. For this reason, the airbag curtain, installed in both sides of the vehicle roof, is a good supplement to the seat belt and the side torso airbag. With a special sensor system that is designed to detect a rollover situation, the airbag could be inflated rapidly enough to prevent head injuries. Because the time for airbag inflation in a rollover situation could be much longer compared to side impacts, it is not a big difficulty to use the side window airbag (e.g., inflatable curtain) for rollover protection also [11-23] (see Figure 11.21).

11.5 Special Requirements and Opportunities for Sensor Applications

For the proper functioning of the pretensioners and airbags, inadvertent ignition must be avoided, and the sensor system must show sufficient performance during accidents. After several years of field experience, knowledge about special effects has grown. In a Ph.D. study [11-24], it was found that the use of several sensors up front, plus central and other sensors used for side impacts, could improve the determination of the severity of an accident. Even side effects such as an increase of signals measured by the sensor accelerometer due to sheet metal oscillations could be observed.

Another effect that has been taken into consideration is optimization of the ignition timing of the airbag system. The following example shows the reason for this statement. In a 40% offset crash against a deformable barrier, the deceleration of the crashed car is not high enough in the beginning of the crash to create a large forward movement of the occupant relative to the vehicle. If the airbag inflates too early and the occupant hits the airbag at a time when the gas in the airbag is already becoming cold and is leaving the airbag through the pop-out valves and the inflator holes, then the occupant's ride downward might be not adequate to prevent high deceleration of the occupant [11-25]. For this reason, the layout of the sensor system must take into consideration the total safety function of the car, including characteristics of the body structure, the sensor system, the pretensioners, and the airbag system. One conclusion of this analysis is that with a shorter ignition time, the inflated airbag must keep the pressure for a longer period of time. This consideration also is important for all types of pre-crash sensors because the pressure in the airbag must be retained over a longer period of time.

Because some automated cruise control (ACC) sensors are used to control the distance between the vehicle and the car in front of it, there is some hope of using these same types of sensors, which are part of accident avoidance systems, for devices in the field of mitigation of injuries. One big European program for the analysis of various sensor systems is Chameleon [11-26]. In this program, not only are different types of sensors used, as shown in Table 11.1, but the possibilities of improving the protection systems in cars for the fulfillment of legal requirements with respect to crash performance are being investigated.

TABLE 11.1
SENSOR SPECIFICATIONS PROVIDED BY
SENSOR SUPPLIERS [11-26]

	Saab	IBEO	TAMAM	Temic	Thales A.S.
Technology	Microwave radar	Laser-rotating radar	Artificial vision	Laser multi-beam radar	Microwave radar
Scan Rate	50 Hz	40 Hz	25 Hz	100 Hz	25 Hz
Delay Time	20 ms	25 ms	40 ms	10 ms	40 ms
Aperture Angle	100°	270°	60°	$3 \times 15°$	60°
Field Depth	0.5–20 m	0.3–20 m	0–40 m	0.5–6 m	0–60 m
Distance Accuracy	0.1 m	0.05 m	3%	0.1 m	1 m, 5%
Angle Accuracy	10°	1°	1°	15°	2°
Velocity Accuracy	5%	±1 kph	6%	10 kph	±0.2 kph

Figure 11.27 shows the possible function of the detection areas of various types of sensors. With an analysis of the combined information, the failure rate with respect to identification of the type of collision partners and the severity of the accident could be reduced significantly [11-27].

One reason for this optimization process is the vision and hope to save lives and reduce severe injuries in the traffic environment. Some future possibilities were demonstrated by DaimlerChrysler at the Frankfurt Motor Show in 2001. In a visionary project for the future called PRE-SAFE, the company mentioned several examples using pre-crash devices, such as the following:

- Early pretensioning of the seat belt
- Earlier ignition of the airbag propellant
- Inclination of declined seat back and positioning of the headrest
- The sunroof will be closed
- The side door panel could be moved closer to the occupant

Although the technology for pre-crash sensor systems will be available relatively soon, its introduction in production cars will depend on the benefit to

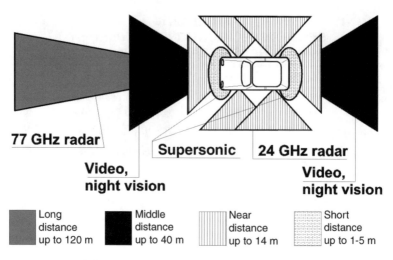

| Long distance up to 120 m | Middle distance up to 40 m | Near distance up to 14 m | Short distance up to 1-5 m |

Figure 11.27 Detection areas of various types of sensors.
(Source: Ref. 11-27.)

the customer and the reliability of the system. One basic requirement should always be kept in mind: The positive effects of new systems should not by themselves create even a single additional failure. Meanwhile, during 2002, the DaimlerChrysler S-Class introduced pre-crash safety items, if a possible accident is defined:

- Pretension of the seatbelts
- The seatback and cushion are brought into a normal position
- The windows and the sunroof are closed

Toyota has announced plans to introduce a pre-crash radar system into production cars in 2003.

11.6 General Literature

- Ziegahn, K.-F. "Fraunhofer Institut Chemische Technologie," Airbag 2000, Fifth International Symposium and Exhibition on Sophisticated Car Occupant Safety Systems, December 4-6, 2000, Karlsruhe, Germany.

- VDI Gesellschaft für Fahrzeug- und Verkehrstechnik: Fahrzeugkonzepte für das 2. Jahrhundert Automobiltechnik, VDI-Berichte 1653, Düsseldorf, 2001, ISBN 3-18-091653-2.

- VDI Gesellschaft für Fahrzeug- und Verkehrstechnik: Innovativer Kfz-Insassen- und Partnerschutz, VDI-Berichte 1637, Düsseldorf, 2001, ISBN 3-18-091637-0.

11.7 References

11-1.   *Automotive Engineering*, No. 4, 2000, Society of Automotive Engineers, Warrendale, PA, United States.

11-2.   Author's information.

11-3.   Audi AG. PROCON-TEN, technical description, 1989.

11-4.   Enßlen, A., et al. "New Safety Elements in Volkswagen-Audi Automobiles," Lecture in Peking, 1996.

11-5.   "Der Neue BMW 7er," *ATZ/MTZ Extra*, November 2001, Vieweg-Verlag, Wiesbaden, Germany.

11-6.   Insurance Institute for Highway Safety status report, Vol. 36, No. 10, November 15, 2001, Arlington, VA, United States.

11-7.   Seiffert, U., et al. "Description of the Volkswagen Restraint Automatic (VW-RA) Used in a Fleet Test Program," SAE Paper No. 740046, Society of Automotive Engineers, Warrendale, PA, United States.

11-8.   States, I.-D., et al. "Volkswagen's Passive Seat Belt/Knee Bolster Restraint (VW-RA): A Preliminary Field Performance Evaluation," Proceedings of the 21st Stapp Car Crash Conference, Society of Automotive Engineers, Warrendale, PA, United States.

11-9.   Holzner, M. "Virtuelle Prozessketten in der Funktionsauslegung des Gesamtfahrzeuges," Proceedings of the *ATZ/MTZ* Congress, July 2002, Berlin, Germany.

11-10. Rabe, M., et al. "Seitenschutz mit Inflatable Curtain und Seiten-Airbag," Conference held by Haus der Technik, March 23–24, 2000, Munich, Germany.

11-11. VW-GIDAS accident database, 2002, MAIS 1+, belted struck side passenger car occupant injuries in side collisions for different severities, multiple body regions possible. Unpublished. Volkswagen AG, Wolfsburg, Germany.

11-12. AUTOLIV GmbH, Elmshorn, Germany, undated. Product information material.

11-13. Temming, J. and Zobel, R. "Frequency and Risk of Cervical Spine Distortion Injuries in Passenger Car Accidents: Significance of Human Factors Data," Proceedings of the International Research Council on the Biomechanics of Impact (IRCOBI) Conference, 1998, pp. 219–234.

11-14. Davidsson, J., Svensson, M.Y., Flogard, A., Haland, Y., Jakobsson, L., Lindner, A., Lövsund, P., and Wiklund, K. "BioRID I—A New Biofidelic Rear Impact Dummy," Proceedings of the International Research Council on the Biomechanics of Impact (IRCOBI) Conference, 1998 (a), pp. 377–390.

11-15. Davidsson, J., Flogard, A., Lövsund, P., and Svensson, M. "BioRID P3—Design and Performance Compared to Hybrid III and Volunteers in Rear Impacts at $\Delta V = 7$ km/h, Proceedings of the 43rd Stapp Car Crash Conference, 1999 (a), pp. 253–365, Society of Automotive Engineers, Warrendale, PA, United States.

11-16. Davidson, J., Lövsund, P., Ono, K., Svensson, M.Y., and Inami, S.A. "Comparison Between Volunteer, BioRID P3, and Hybrid III Performance in Rear Impacts," Proceedings of the International Research

Council on the Biomechanics of Impact (IRCOBI) Conference, 1999 (b), pp. 165–178.

11-17. Cappon, H., Philippens, M., and Wismans, J. "A New Test Method for the Assessment of Neck Injuries in Rear End Collisions," Paper No. 242, Experimental Safety Vehicle (ESV) Conference, Amsterdam, 2001.

11-18. "Head Restraints, Status Report," Vol. 36, No. 9, Insurance Institute for Highway Safety, October 6, 2001, Arlington, VA, United States.

11-19. Saab GmbH. Product information material, undated. Saab Opel Sverige AB, Kista, Sweden.

11-20. Tschachtli, Chr. "Mechanismus gegen Peitschenhieb," *Automobile-Revue* No. 25, June 21, 2001, Bern.

11-21. National Highway Traffic Safety Administration (NHTSA) and Federal Motor Vehicle Safety Standard (FMVSS), FMVSS 216, Roof crush resistance, Washington, DC, United States.

11-22. Rabe, M. "The Role of Glazing for Car Safety," Proceedings of the Glass Processing Days Conference, Tampere, June 6, 2001, Society of Automotive Engineers, Warrendale, PA, United States.

11-23. Meyer, I., et al. "Protection Technology 6/2001, Neues Überschlag-Schutzsystem," *ATZ* 108 (2001) No. 7/8, Wiesbaden, Germany.

11-24. Hübler, R. *Unterstützung bei der Auslegung von Airbagsystemen durch FEM-Berechnungen*, M-Verlag Mainz, 2001, ISBN 3-89653-828-4.

11-25. Schwant, W. "Beurteilung des Schutzpotentials von Airbagsystemen unter Berücksichtigung unterschiedlicher Sensorzeiten," Conference held by Haus der Technik Essen, "Kollisionsschutz im Straßenverkehr," February 29, 2000.

11-26. Fürstenberg, K.Ch. et al. "Development of a Pre-Crash Sensorial System—the CHAMELEON Project," in VDI-Berichte 1653, Vehicle Concepts for the 2nd Century of Automotive Technology, ed. by Verein Deutscher Ingenieure, VDI-Verlag, Düsseldorf, Germany, 2001, pp. 289–310.

11-27. Detlefsen, W. et al. "Evolution of Navigation to Cooperative Traffic," in VDI-Berichte 1653, Vehicle Concepts for the 2nd Century of the Automotive Technology, ed. by Verein Deutscher Ingenieure, VDI-Verlag, Düsseldorf, Gemany, 2001, pp. 357–372.

12.

Pedestrian Protection

12.1 General

Although vehicle safety continues to improve, it is evident that if only passenger car occupants become safer in the traffic environment, the total necessary reduction of fatalities and severely injured people cannot be achieved. This means that other groups, such as pedestrians, must be considered. If we compare the situation in Europe in 1980, we could identify approximately 15,000 fatalities in pedestrian accidents, whereas in 1998 this number was reduced below 7,000 fatalities [12-1], as shown in Figure 12.1.

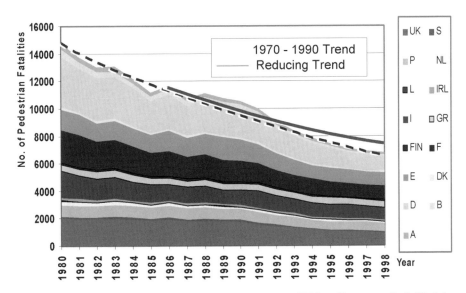

Figure 12.1 Fatal pedestrian injuries in Europe (EU). (Source: Ref. 12-1.)

The percentage of pedestrian fatalities with respect to all accidents in Germany was 23.7% (3,095 out of 13,041) in 1980, and 12.6% (977 out of 7,761) in 1998. This means that not only the absolute number but the percentage of fatal pedestrian accidents was reduced [12-2]. From an analysis of accident data in Germany, we can identify that the total number of injuries, serious injuries, and fatalities is highest in urban traffic, with a total of 37,581 (93.2%). This is followed by rural accidents with a total of 2,539 (6.3%), and on motorways, with a total of 185 (0.5%). If we examine the number of fatal accidents, that number is 983, with 67% in the city, 29% rural, and 4% on motorways. Figure 12.2 shows these results in more detail.

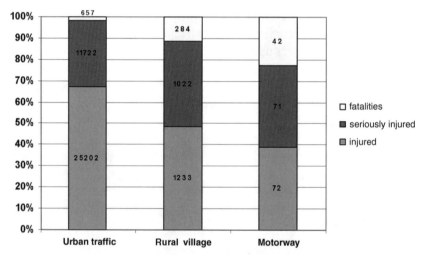

Figure 12.2 Pedestrian accidents, as a function of road type.
(Source: Ref. 12-2.)

The reduction of fatalities was achieved by vehicle improvements and other measures in the traffic environment. Figure 12.3 supports that vehicle design definitely has had a positive influence on safety. In Figure 12.3, we find the frequency of pelvis/femur fractures for pedestrians older than 11 years old [12-3] and a comparison between older cars (1974 to 1983) and newer cars (from 1989). The improvements can be stated as being significant.

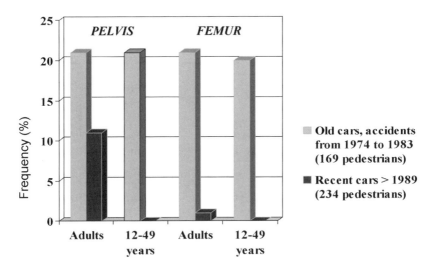

Figure 12.3 Frequency of pelvis/femur fractures of pedestrians
older than 11 years. (Source: Ref. 12-3.)

Figure 12.4 shows how complicated the accident situation can be in the real world. This graph shows the head contacts on the front of the vehicle, as a function of injury severity and collision severity.

There is a high distribution of contact areas on the vehicle in relation to the Abbreviated Injury Scale (AIS). An in-depth study conducted by a university team analyzed pedestrian accidents in the city of Hanover since 1985, with respect to injury frequency to various body regions [12-4]. Figure 12.5 shows the results of the analysis.

This figure can be interpreted as follows: Head injuries occurred with 32.5%, upper extremities 15.6%, and lower extremities 32.6%. The regions of the front hood contributing to the injuries are the windscreen and its frame 28.5%, the bonnet for head injuries 17.3%, the leading edge of the bonnet to hip injuries 40.1%, and leg injuries by the bumper with 44.2%. If we analyze all injuries related to vehicle parts, the distribution is as follows: 15.6% by the windshield and its frame, 13.3% from the bonnet/fenders, 6.9% from the leading edge of the bonnet, and 15.3% from the bumper.

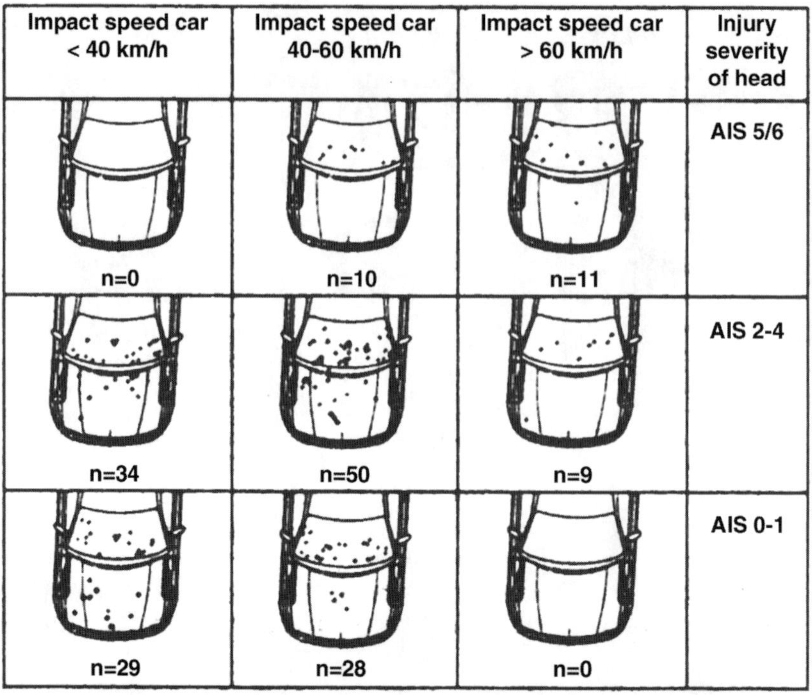

Impact speed car < 40 km/h
Impact speed car 40-60 km/h
Impact speed car > 60 km/h
Injury severity of head
n=0
n=10
n=11
AIS 5/6
n=34
n=50
n=9
AIS 2-4
n=29
n=28
n=0
AIS 0-1

Figure 12.4 Head contact of pedestrians on the front hood of the vehicle, versus injury severity and collision speed. (Source: Ref. 12-1.)

32.5%
28.5%
1.1%
17.3%
8.5%
15.6%
15.6%
2.8%
13.3%
6.0%
40.1%
6.9%
4.1%
44.2%
32.6%
9.8%
15.3%
15.7%

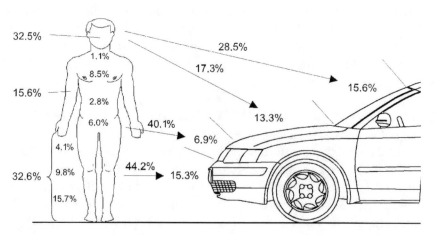

Figure 12.5 Injury frequency to various body regions of pedestrian and vehicle collisions. (Source: Ref. 12-4.)

Also, contact with the road surface as a secondary impact causes different severe injuries. Therefore, further improvements cannot be achieved by single measures but only by the sum of several activities. Because of the mixed traffic on the roads particularly in Europe and Asia—a large number of cars, two-wheelers, and pedestrians—pedestrian protection has received great attention from the public and, as a consequence, by legal authorities. Realizing that, in many developing countries we have a dramatic change in traffic participants. Thus, pedestrian protection might become even more important. Many actions in the field of accident avoidance for pedestrians were mentioned in Chapter 6. Future improvements for pedestrian protection must consider the possibilities in the areas of both accident avoidance and mitigation of injuries. Figure 12.6 describes the task of the driver. Often, the critical situation of a potential accident is underestimated by the driver. Likewise, pedestrians, especially children, do not pay enough attention to other traffic participants.

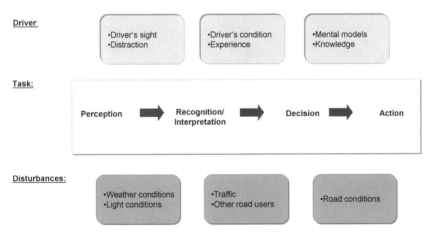

Figure 12.6 The task of the driver. (Source: Ref. 12-2.)

In the chapter on accident avoidance with respect to the reduction of collisions between pedestrians and other traffic participants, several measures were recommended. The most important parameters are as follows:

- View of the driver, also to avoid blind spots (a priority for truck drivers)

- Detection of pedestrians, especially in dark environments

- Traffic management and road design

- Traffic signal improvement

- Optimization of the red-light time on crosswalks for pedestrians

- Measures on the vehicle, such as antilock brake systems (ABS) and day-time running lights

The area (i.e., accident avoidance or mitigation of injuries) in which the level of protection achieved is greater can be determined only in the future. We also should keep in mind that the task is complex due to the different ages and sizes of pedestrians. Directly related to age is the physical resistance of the pedestrian; that is, older people have a higher risk of bone fractures.

The knowledge that pedestrian protection in Europe is of great importance has stimulated intense research activities during recent years. The EEVC with its Working Group 17 [12-5] finalized scientific work at the end of 1998. In this work, a good balance between test requirements and real injury risk, new biomechanical knowledge, and the development of test experience, EEVC has created the basis for all current activities. For example, the proposal introduced by the European Commission [12-6] and information to the consumer provided by the European NCAP [12-7] are based on EEVC activities. It has been some time since the NCAP findings presented the first results in the area of pedestrian protection also. Although there is no direct advantage for the vehicle owner and the self-protection of the car occupant, the public interest related to pedestrian protection is increasing. As an example, a report in a German automotive magazine stated that the Honda Civic is a very safe passenger car with respect to pedestrian protection by detail optimization of the vehicle. For example, the suffi-cient distance between the front hood and the rigid parts below it could contribute to reduction of injuries, particularly for children.

12.2 Requirements by the European Commission

The detailed requirements of the European Commission are defined in Annex II, Parts A and B, of the commission paper on pedestrian protection [12-6]:

Part A. In the lower legform to bumper tests at impact speeds of up to 40 km/h [25 mph], the maximum dynamic knee bending angle shall not exceed 21.0°, the maximum dynamic knee shearing displacement shall not exceed 6.0 mm [0.24 in.], and the acceleration measured at the upper end of the tibia shall not exceed 200g.

In upper legform to bonnet leading edge tests at impact speeds of up to 40 km/h [25 mph], depending on the shape of the vehicle, the instantaneous sum of the impact forces with respect to time, to the top and bottom of the impactor, shall be recorded and compared with the possible target of 5.0 kN, and the bending moment on the impactor, at any of the three measuring positions, shall be recorded and compared with the possible target of 300 Nm. This test is defined as for monitoring purposes.

In the upper legform to bumper tests at impact speeds of 40 km/h [25 mph], the instantaneous sum of the impact forces with respect to time, to the top and bottom of the impactor, shall not exceed 7.5 kN, and the bending moment on the impactor, at any of the three measuring positions, shall not exceed 510 Nm.

In child headform (3.5-kg [7.7-lb] impactor) to bonnet top test at impact speeds of 35 km/h [22 mph], the head performance criterion (HPC), calculated from the resultant of the headform accelerometer time histories, in accordance with Paragraph 2.5 of Annex I, shall not exceed 1000 for 2/3 of the bonnet test area, 2000 for 1/3 of the bonnet test area as defined in Paragraph V 3.2 of Annex III.

In adult headform (4.8-kg [10.6-lb] impactor) to windscreen test at impact speeds of 35 km/h [22 mph], the head performance criterion (HPC), calculated from the resultant of the headform accelerometer time histories, in accordance with Paragraph 2.5 of Annex I, shall not exceed 1000. This test is defined as for monitoring purposes.

Part B. In the lower legform to bumper tests at impact speeds of 40 km/h [25 mph], the maximum dynamic knee bending angle shall not exceed 15.0°, the maximum dynamic knee shearing displacement shall not exceed 6.0 mm [0.24 in.], and the acceleration measured at the upper end of the tibia shall not exceed 150g.

In upper legform to bumper tests at impact speeds of 40 km/h [25 mph], the instantaneous sum of the impact forces with respect to time, to the top and bottom of the impactor, shall not exceed 5.0 kN, and the bending moment on the impactor, at any of the three measuring positions, shall not exceed 300 Nm.

In upper legform to bonnet leading edge tests at impact speeds up to 40 km/h [25 mph], depending on the shape of the vehicle, the instantaneous sum of the impact forces with respect to time, to the top and bottom of the impactor, shall not exceed 5.0 kN, and the bending moment on the impactor, at any of the three measuring positions, shall not exceed 300 Nm.

In adult (4.8-kg [10.6-lb] impactor) and child headform (2.5-kg [5.5-lb] impactor) to bonnet top tests at impact speeds of 40 km/h [25 mph], the head performance criterion (HPC), calculated from the resultant of the headform accelerometer time histories, in accordance with Paragraph 2.5 of Annex I, shall not exceed 1000 on the whole bonnet test area.

Figure 12.7 shows a more general view of the requirements of the European Commission.

There are many more details to be observed. Some examples are shown in the following descriptions:

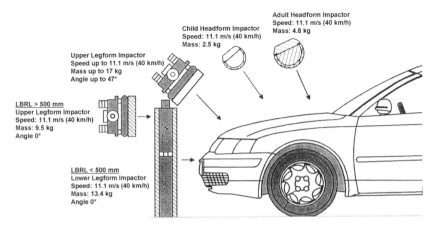

Figure 12.7 Requirements for pedestrian protection.
(Source: Ref. 12-7.)

- The lower legform test is performed to check the bumper and the front end of the vehicle. The impact speed is 40 km/h (25 mph), and the dynamic knee bending angle shall not exceed 21° from the vertical. The shearing displacement shall not exceed 6 mm (0.24 in.), and the acceleration measured at the upper end of the tibia shall be below 200g.

- The velocity of the upper legform impactor (see the top of Figure 12.8), depends on the shape of the vehicle (see the bottom of Figure 12.8).

- Also, the angle of the upper legform impact to the bonnet edge depends on the vehicle shape.

- The child headform top test differentiates the zone where the head performance criterion (HPC) does not exceed 1000 in Zone A or 2000 in Zone B (see Figure 12.9).

- The head impact form has a weight of 3.5 ± 0.07 kg $(7.7 \pm 0.15$ lb). The impact speed is 35 km/h (22 mph).

- The adult headform should impact the windshield at a speed of 35 km/h (22 mph), whereas the head performance criterion (HPC) should be below 1000.

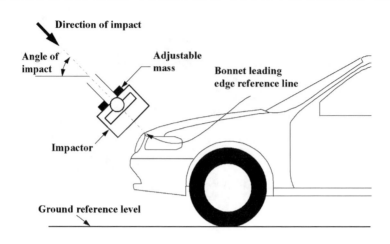

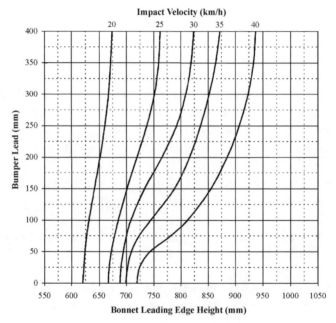

Figure 12.8 Bonnet leading edge test procedure.
(Source: Ref. 12-6.)

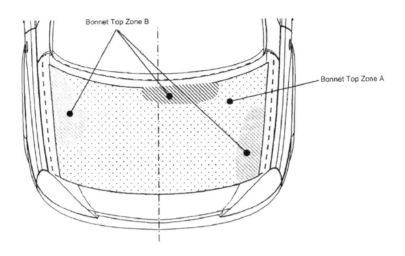

Figure 12.9 Bonnet head impact zones. (Source: Ref. 12-6.)

- Figure 12.10 shows the adult headform and child headform impactor.

- Both are applicable to Part B of the commitment. The weight of the adult headform is 4.8 ± 0.1 kg (10.6 ± 0.22 lb), and 2.5 ± 0.05 kg (5.5 ± 0.11 lb) for the child head impactor. The impact area is determined by the wrap-around length, which for the child head impact is between 1000 and 1500 mm (39 and 59 in.), and for the adult head impact is between 1500 to 2100 mm (59 and 83 in.).

12.3 The Legal Situation

Until recently, few legal requirements were enforced with respect to pedestrian protection. One of these is EG Directive 74/483 EWG and ECE 26 [12-8], which covers external projections. Because of the complexity of future European requirements, similar to the voluntarily commitment by ACEA (The European Automobile Manufacturers Association) to reduce CO_2 emissions, the commission of European communities has reached an agreement with ACEA related to pedestrian protection. The basic points are described in the final document published by the commission on July 11 COM (2001)

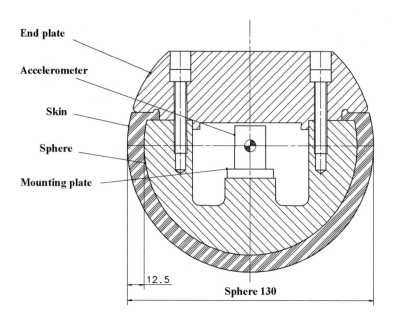

End plate
Accelerometer
Skin
Sphere
Mounting plate

12.5

Sphere 130

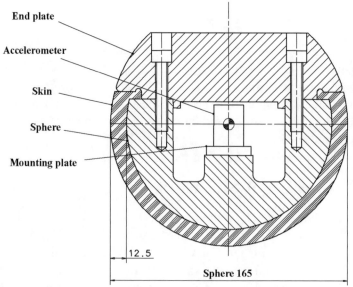

End plate
Accelerometer
Skin
Sphere
Mounting plate

12.5

Sphere 165

Figure 12.10 Child and adult headform impactors.
(Source: Ref. 12-6.)

389 titled "Pedestrian Protection: Commitment by the European Automobile Industry."

The main elements of the commitment are as follows:

- Industry commits to meet the recommendations of the Commission Joint Research Center as a first step with the passive safety measures, with all new types of vehicles fulfilling the requirements from July 1, 2005; 80% of all new vehicles from July 1, 2010; 90% in 2011; and the remaining 10% by 2012.

- Industry commits to equip all new vehicles with antilock brake systems (ABS) in 2003 and daytime running lights (DRL) beginning in 2003 for new vehicles.

- Rigid bull bars will not be installed on new vehicles after 2002.

- Industry supports the objectives of the e-Europe Action Plan and agrees on the importance of additional ICT elements in improving active safety (an indicative list is annexed to the commitment), and it commits to install progressively additional active safety devices on all new motor vehicles.

- Compliance with the European Enhanced Safety Vehicle Committee (EEVC) targets for pedestrian safety for all new types of vehicles in 2010 and for all new vehicles progressively from 2012, but not later than the end of 2014, through EEVC technical prescriptions or other measures that are at least equivalent (at least equal protective effects), subject to a feasibility assessment. This assessment will be undertaken by July 1, 2004, by a Monitoring Committee set up by the Commission, with industry participation. The assessment will be based on the findings of independent bodies, and it also will cover the date of application of the EEVC requirements for those vehicles that, in exceptional cases, might not be able to meet the EEVC requirements by the established dates.

- Compliance reporting as well as reporting on technical progress and planning will be submitted to the Monitoring Committee. All technical compliance verification will be conducted by independent technical services.

In addition, the commitment includes several general provisions and, in particular, the following:

- The automobile industry supports future efforts to achieve international harmonization in the area of pedestrian protection, in the framework of the UN/ECE 1998 Agreement on the establishment of global technical regulations.

- Compliance with EC competition rules: The commitment will be implemented in compliance with EC competition rules. As far as the commitment would contain certain restrictions of competition, a formal notification could be transmitted to the Commission, provided that this possibility is offered by applicable EC competition rules.

Commission Recommendation. As an additional security for ACEA to respect its commitment, the Commission will make it clear in its recommendation that it will consider regulatory measures should ACEA not honor its commitment.

Evaluation of the Commitment. In December 2000, the Commission agreed on elements to be met by an industry commitment on pedestrian safety.

The Commission has assessed the ACEA commitment in light of these previously mentioned elements as follows:

ACEA members have agreed to meet a high level of pedestrian protection, by meeting the EEVC requirements or by introducing measures offering at least equivalent protective effects in 2010. As a first step in this direction, the industry has agreed, on one hand, to meet the requirements proposed by the JRC in its report of December 19, 2000, from July 1, 2005. On the other hand, the industry has agreed to supplement this measure with three additional initiatives also conducive to improved pedestrian and road safety—that is, the equipment of all new motor vehicles with antilock brake systems (ABS) and daytime running lights (DRL) and the end of sales of rigid bull bars by car manufacturers from 2002.

According to the commitment, industry will meet a first significant package of pedestrian safety measures by July 1, 2005. However, an important package of additional safety measures will already be introduced in the period 2002 to 2004 (see preceding paragraph).

This proposal must be agreed on by the European parliament, which planned to discuss this in 2002. Table 12.1 [12-9] shows a comparison of the EEVC and ACEA. Without the fundamental work of the European Enhanced Vehicle Safety Committee, the necessary knowledge for performance and test criteria would not exist.

12.4 Technical Solutions for Vehicles in Accidents

The design trend for vehicles with respect to aerodynamic optimization, styling elements, and technical changes such as plastic front ends, including the bumper, definitely helped to reduce the negative consequences in pedestrian accidents. Also, the facts that the outside contour of the vehicle must have only parts with a radius greater than 3.2 mm (0.13 in.) and that the outside mirrors must fold away during an impact have reduced the number of injuries. This also is true for the glued-in laminated windshield, which prevents head injuries to a greater extent, especially cuts from glass separation.

The new European requirements are demanding many more changes for vehicles. Chapter 13 discusses some examples of possible solutions. It is evident that three areas in particular must be redesigned:

- During the leg test, the front impact zone with its bumper system must absorb the kinetic energy of the pedestrian leg impact device and must prevent the rotation of the legform. Whereas the impact energies can be reduced due to energy-absorbing components in the front bumper, the rotation of the leg must be reduced by the geometric shape of the front of the vehicle.

- The hip impact test requires a redesign of the front hood, with the front hood latch transverse bar.

TABLE 12.1
COMPARISON OF RECOMMENDATIONS
IN TERMS OF CONTENTS OF TEST PROCEDURE
FOR PEDESTRIAN PROTECTION (EEVC, ACEA)

Head Impact Device	EEVC WG 17 Child/Adult	ACEA Part A Child/Adult
Impact Angle [°]	50/65	50/65 monitoring purposes
Impact Velocity [m/s]	11.1 ± 0.2	9.7 ± 0.2
Mass [kg]	2.5/4.8	2.5/4.8 monitoring purposes
Test Criteria, HIC	<1000	1/3 <2000 2/3 <1000 bonnet top

Hip Impactor	EEVC WG 17	ACEA Part A Monitoring Purposes
Impact Angle [°]	10–47	10–47
Impact Velocity [m/s]	5.6–11.1 ± 2%	5.6–11.1 ± 2%
Mass [kg]	9.5–18	9.5–18
Test Criteria: Force	<5 kN	<5 kN
Torque	<300 Nm	<300 Nm

Leg Impact Device	EEVC WG 17	ACEA Part A Monitoring Purposes
Impact Angle [°]	0 ± 2	0 ± 2
Impact Velocity [m/s]	11.1 ± 0.2	11.1 ± 0.2
Mass [kg]	13.4	13.4
Test Criteria: Angle	<15	<21°
Shear	<6 m	<6 mm
Acceleration	<150g	<200g

- The most complicated part of the requirements is the reduction of the deceleration level during the headform impact to fulfill the head protection criteria (HPC) at the front hood of the vehicle. If we assume that the impact speed is 40 km/h (25 mph) and the average deceleration level should not exceed 60g, then we already need a free deformation length below the front hood of approximately 100 mm (4 in.).

In Figure 12.11, computer animation shows which vehicle components under the front hood would have to be redesigned [12-1].

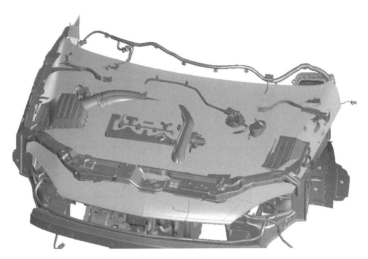

Figure 12.11 Engine compartment with an 80-mm (3.1-in.) lowered front hood. (Source: Ref. 12-1.)

Because of the design restrictions and the location of vehicle components under the front hood and to accommodate the requested free field of view for the driver, different design solutions are being investigated to ensure that the front hood is not built too high. For example, if a pedestrian hits the bumper, a mechanical or pyrotechnic upward movement of the front hood could achieve the necessary free deformation area. Another solution was demonstrated first at the International Automobile Show in Frankfurt [12-10]. Figure 12.12 shows possible improvements in one specific vehicle, which were achieved by dynamic uplifting of the front hood rear edge by approximately 100 mm (4 in.) and the use of pedestrian protection airbags.

Via two airbags on the left and the right sides at the rear part of the front hood it is lifted upward. The bags also improve the head impact situation against the windshield. For all systems that must be activated by pyrotechnic actuators or airbags, the sensor system must discriminate not only objects of

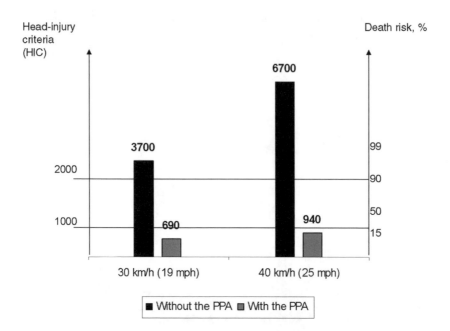

Head-injury
criteria
(HIC)

Death risk, %

6700

3700

2000 ────────────────────────────── 99
                                    90

1000 ───── 690 ──────── 940 ───── 50
                                    15

30 km/h (19 mph)   40 km/h (25 mph)

■ Without the PPA ■ With the PPA

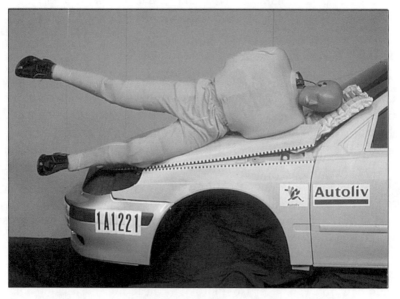

Figure 12.12 Pedestrian protection via front hood airbags.

different sizes but objects of different stiffnesses as well. This means, for example, that the sensor system must be able to differentiate an impact against pedestrians or trees. It also is important that at low-speed impacts against obstacles other than pedestrians, the system will not be activated. For this reason, the priority lies in the development of reliable sensor systems.

12.5 References

12-1. Hahn, W. "Fußgängerschutz—Von der Unfallanalyse zur Entwicklungsanforderung," Verband der Automobilindustrie (Association of the [German] Automobile Industry) (VDA) Technical Congress, 2001, Frankfurt am Main.

12-2. Hartlieb, M. "Fußgängerschutz-Möglichkeiten und Potenziale zukünftiger Technologien," Verband der Automobilindustrie (Association of the [German] Automobile Industry) (VDA) Technical Congress, 2001, Frankfurt am Main.

12-3. INRETS (Institut National de Recherche Sur Transports et Leur Securite), Pedestrian Accident Analysis, Diagnosis and Recommendations, France, 1999.

12-4. Otte, D. "Severity and Mechanism of Head Impacts in Car to Pedestrian Accidents," International Research Council on the Biomechanics of Impact (IRCOBI) Conference, Sitges, 1999.

12-5. European Enhanced Vehicle Safety Committee (EEVC), Working Group 17, "Final Report on Improved Test Methods to Evaluate Pedestrian Protection Afforded by Passenger Cars," December 1998.

12-6. Commission of the European Communities, "Communication from the Commission to the Council and the European Parliament." Pedestrian Protection: Commitment by the European Automobile Industry, Brussels, July 2001. European Enhanced Vehicle Safety Committee (EEVC), Working Group 17, "Final Report on Improved Test Methods to Evaluate Pedestrian Protection Afforded by Passenger Cars," 1998.

12-7.   Kallikske, I. "Requirements for Pedestrian Protection at Motor Vehicles and Solution Concepts," Proceedings of the Verein Deutscher Ingenieure VDI-K Congress 2001. Also Friesen, F., et al. "Optimierung von Fahrzeugen hinsichtlich des Beinaufpralltests," *ATZ* Vol. 104, No. 5, 2002, ISSN 0001-2785, Vieweg-Verlag, Wiesbaden, Germany. Also European New Car Assessment Program (EURONCAP), "New Testing and Assessment Protocol" (V 3.1, November 2001).

12-8.   74/483/EWG and ECE R 26, External Projections.

12-9.   Ros, E. "Neue Sicherheitsentwicklungen für Geländewagen," Proceedings 3, Grazer Allradkongress, Graz, 2002.

12-10.   AUTOLIV GmbH. Press release, Internationale Automobil-Ausstellung (IAA) (International Automobile Exhibition), Frankfurt am Main, 2001.

13.
Compatibility

Because of progress in the field of mitigation of injuries for vehicle occupants, more activities in areas other than the simulation of classic accidents (such as front, lateral, and rear impacts) have been investigated in greater detail. One of these activities is the question of compatibility among traffic participants and how that compatibility could be increased. Compatibility in this context means the positive performance of traffic participants among each other in the event of an accident. In the traffic environment, the following group of participants must be considered: pedestrians, two-wheelers, passenger cars, trucks, and occupants of buses. Injuries could occur during impact with another collision partner or during a single accident against a fixed obstacle. For compatibility investigations, it has been shown that a global approach, which includes all groups of possible collision partners, is complicated, and the solutions are more than complex. Thus, in the following analysis, this compatibility view is limited to passenger car collisions and passenger car occupants.

For the layout of the vehicle design, the following criteria have a direct influence on the collision process:

- Mass ratio of the collision partners

- Geometry of the vehicle structure

- Force–deflection characteristic of the vehicle structure

- The location of the installation, size, and mass of the powertrain unit

- The type and rigidity offered by the occupant cell (survival space)

- The performance of the steering wheel and column, dashboard, knee impact zone, and pedals during deformation

Research activities on the subject of compatibility were performed as early as 1970 [13-1]. Because of other priorities, the intensity of research on compatibility was relatively low but increased during the mid-1990s. Before the theoretical analyses are discussed, we first should examine the relevant accident data. Figure 13.1 shows the distribution of collision partners for passenger car occupants in frontal collisions in Western Europe.

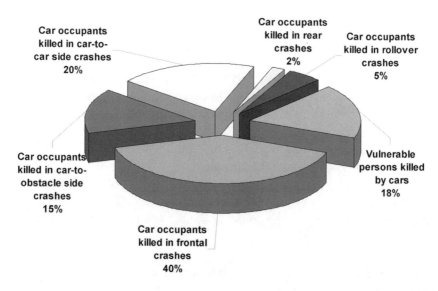

Figure 13.1 Collision distribution for fatal injuries in the European Union. (Source: Ref. 13-2.)

As can be seen from the database, approximately 40% of passenger car accidents with fatalities are frontal collisions, followed by lateral collisions. If we analyze the European situation, Figure 13.2 shows the distribution of severe injuries and fatalities for several countries.

If we examine the group with severe injuries, car-to-car collisions in most European countries has the highest priority. Figure 13.3 shows that for

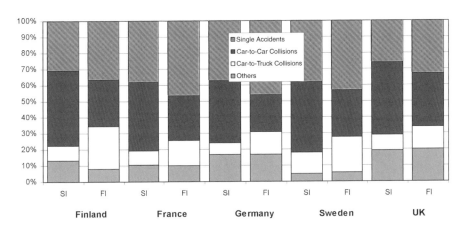

Figure 13.2 Collision distribution for fatalities and severe injuries in some European countries. (Source: Ref. 13-3.)

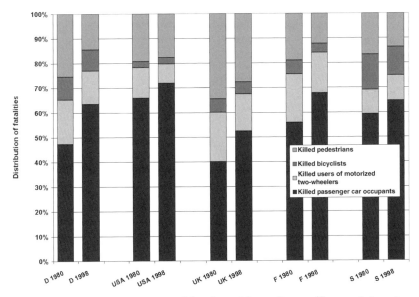

Figure 13.3 Comparison of fatal accidents for traffic participants in 1980 and 1998. (Source: Ref. 13-4.)

passenger car occupants, although the absolute numbers were reduced, the totals are approximately 50% in the United Kingdom and up to 70% in the United States with respect to fatal accidents.

243

Although the relative number of fatal accidents for vehicle occupants decreased significantly during this period, the relative number increased. The highest rate was in the United States, with more than 70%; the lowest rate was in the United Kingdom, with 52%. In addition to the type of accident, the vehicle mass involved is one of the most important parameters. Because vehicle mass, even in the subcompact and compact areas, has increased over the years and the upper-car-class weight has not increased to the same extent, we find that 90% of accidents have a maximum difference in vehicle mass of a factor of 1.6. Figure 13.4 illustrates the sum frequency of colliding vehicles as a function of mass ratio. One interesting observation could be made. From an analysis in 1974, it could be seen that this mass ratio covered 85% of all passenger-to-passenger car collisions in Europe and more than 80% in the United States. This shows that despite the increase in absolute weight, the weight difference did not increase to the same extent.

Another important element is the relative velocity of vehicles with severe injuries (MAIS 3+). Figure 13.5 shows that a $v_{relative}$ of approximately 170 km/h (106 mph) again covers 90% of all accidents. With respect to the

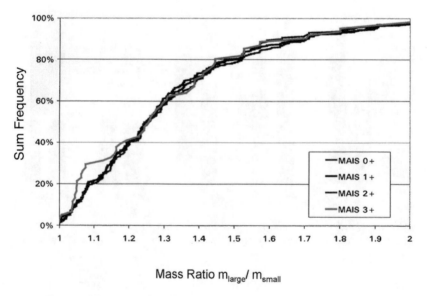

Figure 13.4 Accident frequency as a function of mass ratio.
(Source: Ref. 13-5.)

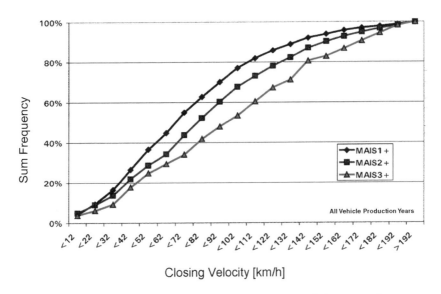

Closing Velocity [km/h]

Figure 13.5 Accident frequency as a function of closing velocity.
(Source: Ref. 13-5.)

overlapping ratio of the vehicle, the accident analysis data show that a high percentage of accidents occurs with an overlapping of less than 50%. Also, the deformation length of the vehicles in relation to severe and fatal injuries shows a direct correlation. If cars are so heavily deformed that the deformation already reaches the windshield area, the probability of severe injuries is increasing.

13.1 Theoretical Analysis

As mentioned, the legislation today mainly covers self protection, which means the protection of the occupants in their own vehicles. If we compare two vehicles where m_1 = 1000 kg (2205 lb) and m_2 = 1600 kg (3527 lb) in a vehicle against a fixed barrier collision and use triangular force deflection characteristics, then we find a deformation length of 0.6 m (2 ft) in a 50-km/h (31-mph) frontal barrier crash (plastic deformation). Figure 13.6 shows the force deflection characteristics.

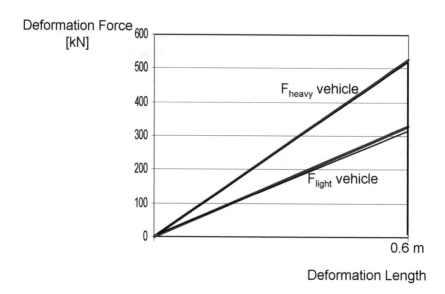

Figure 13.6 Theoretical force deflection characteristics in a crash against a fixed barrier.

The maximum deformation force can be calculated by multiplying the mass times the maximum deceleration, neglecting that a partial portion of the vehicle mass is already at a standstill. The maximum deceleration is 33g, and the maximum forces for the vehicle m_1 with 1000 kg (2205 lb) are 330 kN; for the vehicle m_2 with 1600 kg (3527 lb), the forces increase to 528 kN. Figure 13.7 illustrates the car-to-car collision, where two vehicles collide against each other. The change of velocity can be calculated by the formulas

$$\Delta v_1 = v_r \cdot \frac{m_2}{m_1 + m_2}$$

and

$$\Delta v_2 \cdot \frac{m_1}{m_1 + m_2}$$

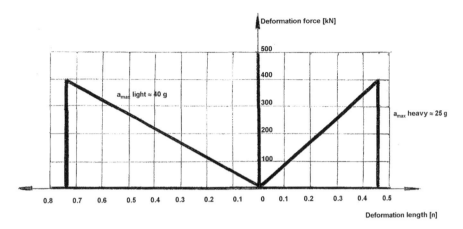

*Figure 13.7 Theoretical force deflection characteristic,
as a function of deformation length.*

$$\Delta v_1 = 100 \text{ km/h} \cdot \frac{1600 \text{ kg}}{2600 \text{ kg}} = 61.50 \text{ km/h}$$

and

$$\Delta v_2 = 100 \text{ km/h} \cdot \frac{1000 \text{ kg}}{2600 \text{ kg}} = 38.46 \text{ km/h}$$

The deformation length of the lighter vehicle increases with this theoretical observation to $\cong 0.74$ m (2.43 ft). For the heavier vehicle, the deformation length is reduced to approximately 0.46 m (1.5 ft). From this theoretical example, we can easily follow the conclusion that this deformation must be available in the light vehicle without disturbing the occupant cell. Both the larger deformation and the higher g-level in the lighter vehicle must be considered with respect of the layout of the restraint system.

The question of whether a compatible free-deformation characteristic could be achieved for various vehicles is not answered easily. In a study performed by Huibers [13-6], the force deflection characteristics for a number of cars measured behind the deformable barrier element demonstrates interesting results. Figure 13.8 shows the average force deflection characteristic for various groups of cars. The four different vehicle groups (phases) of the force/deflection characteristic are shown. Number 3 shows subcompact cars sold in Europe, such as the Volkswagen Golf or the Toyota Corolla. Number 4 shows larger vehicles, such as the Audi A6 or the Mercedes E 200. Number 6 shows MPVs, such as the Volkswagen Sharan or the Chrysler Voyager. Number 7 shows minis, such as the Volkswagen Polo or the Peugeot 206.

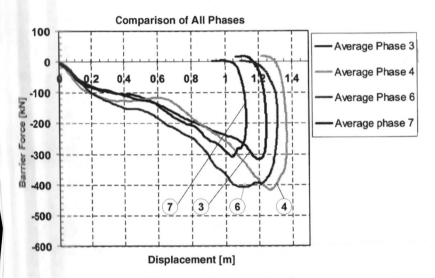

Figure 13.8 Force deflection characteristics for various vehicle groups.
(Source: Ref. 13-6.)

It is interesting that for all passenger cars to a deformation length (offset test) of 1 m (3.3 ft), the force deflection characteristic is very similar. This is not the case for the MPVs. We believe this situation can be solved. This requires a test or a computer simulation that eliminates the existing differences of today. Keeping the preceding results in mind, we should concentrate on the geometric layout of the vehicle structure and the strong integrity of the vehicle compartment.

Table 13.1 provides an overview of various activities related to research and actual proposals for the judgment of the compatibility behavior of passenger cars.

TABLE 13.1

PROPOSALS FOR THE EVALUATION OF COMPATIBILITY

Test Procedure	Cell Resistance	Force Aggressiveness		Self Protection		Structural Aggressiveness				Difference Barrier and Veh/Veh Crash
		Front-End Resistance	Deceleration Pulse as Veh/Veh Crash	Cell Deformation	Occupant Loading	Force Distribution	Prevention of "Bottoming Out"	Force Level	Shear Force	
TRL Force Measurement Wall 100%	No	No	No	No	Yes	Yes	No	Yes	No	No
TRL Force Measurement Wall 40%	No	No	No	Yes	Yes	Yes	No	Yes	No	No
MIRA Force Measurement Wall	No	No	No	No	Yes	Yes	No	Yes	No	No
U.S. NCAP	No	No	No	No	Yes	Yes	No	Yes	No	No
ADAC Barrier	No	No	No	Yes	Yes	`No	Yes	No	No	No
Progressive Deformable Barrier (PDB)	No	Yes	No	No	No	Yes	Yes	Yes	Yes	No
NHTSA Moving Deformable Barrier (MDB)	No	No	Possible	Yes	Yes	Possible	Depends on deformation element	Possible	Depends on deformation element	Possible
Hydraulic Deformable Barrier (HDB)	Yes	Yes	No	Yes	Yes	Yes	Yes	Yes	Possible	Possible

One new approach was described by Schwarz and Zobel [13-1]. They define a bulkhead concept that requests a minimum resistance of the occupant cell, together with a test for the agressiveness of the front end of a vehicle. The front end could be checked, for example, in a crash against a special hydraulic deformable barrier, where the energy absorption capability of the impacting car versus this barrier could be used as a scale for compatibility.

This approach also helps in the design phase of a vehicle. Relou [13-7] has developed a simplified mathematical model, where he simulates the front structure by a mathematical model. This model is much easier to handle with respect to computer power, cost, and time than the finite element model. The different force deflection zones are reflected versus an optimal deformation force distribution of the front structure of the vehicle. Figure 13.9 shows the layout of the model.

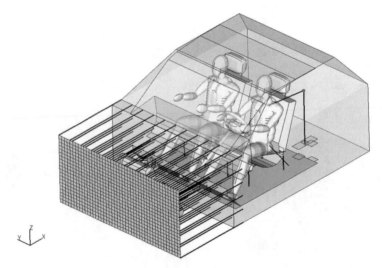

Figure 13.9 Simulation model for frontal impacts.
(Source: Ref. 13-7.)

In Figure 13.10, the actual data of the investigated vehicle are shown versus the optimal compatibility index [KVG], which is a pre-defined force deflection characteristic. The left side of the figure shows the actual force versus deformation, and the right side shows the compatibility index (KVG).

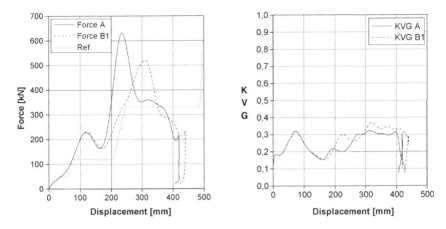

Figure 13.10 Compatibility index versus deformation.
(Source: Ref. 13-7.)

The advantage of this mathematical approach is the possibility of optimizing by computer techniques the different vehicles of one car manufacturer or of a group such as Ford, Volkswagen, or DaimlerChrysler, without performing expensive car-to-car or car-to-barrier crash tests. It is evident that this method also could be used to simulate lateral collisions. With these new tools, it is possible to simulate a great portion of the real-world accident science and to use this data for vehicle design.

The geometric structure of the vehicles also is very important. This becomes even more important if we examine sport utility vehicles (SUVs) and light-duty trucks, as well as collisions between passenger cars and trucks. For compatibility between SUVs and passenger cars, a redesign of the front end of the SUV is necessary. This redesign must develop the possibility that the structural elements of both vehicles could match in a collision. For the truck design, Schimmelpfenning developed an interesting approach [13-8]. Figure 13.11 shows an aluminum frame around the lower part of the trailer of a truck. This device is already in production in small numbers by a trailer manufacturer in Germany.

We can draw some conclusions from the preceding information. The compatibility of collision partners in real-world accidents is receiving increasing

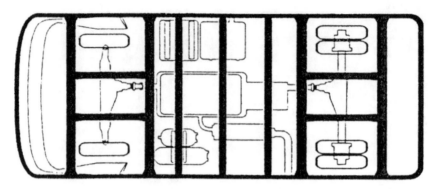

Figure 13.11 Aluminum board frame for a truck. (Source: Ref. 13-8.)

attention. In the priority for the future rule-making process, it is important not to further increase self protection while reducing the protection of collision partners. For example, because of a higher impact speed against a frontal barrier, if the deformation force of the relevant vehicle must be increased, the vehicle might become too aggressive. The other method, which is to make the crash length of this relevant vehicle longer, is impossible because of space and weight factors. Whether the described method of a bulkhead layout and a computerized model, or tests with real cars against a type of deformable barrier, can be introduced in the rule-making process will be determined in the future.

13.2 References

13-1. Appel, H. and Deter, T. "Crash Compatibility for Passenger Cars— How to Achieve?" in VDI-Berichte 1471, Innovativer Kfz-Insassen- und Partnerschutz, ed. by Verein Deutscher Ingenieure, VDI-Verlag, Düsseldorf, Germany, 1999.

See also:

Appel, H. "Sind kleinere Wagen unsicherer als große? Unfallhäufigkeit, Selbstschutz und Partnerschutz bestimmen die Sicherheit," VDI-Nachrichten, 1975, No. 7, pp. 14–16.

Bangemann, C. "Knall-Studie," *Auto, Motor, Sport*, 1999, No.19, pp. 42–49.

Brite Euram 4049: The Development of Criteria and Standards for Vehicle Compatibility, Final Report 2001.

Digges, K. et. al. Stiffness and Geometric Incompatibility in Collisions Between Cars and Light Trucks, SAE Paper No. 2003-01-0907, Society of Automotive Engineers, Warrendale, PA, United States.

Edwards, M., et al. "Compatibility—The Essential Requirements for Cars in Frontal Impacts," International Conference on Vehicle Safety 2000, Institution of Mechanical Engineers, London, UK, June 7–9, 2000.

Färber, E. "EECV Research in the Field of Improvement of Crash Compatibility Between Passenger Cars," 17th International Technical Conference on the Enhanced Safety of Vehicles, Amsterdam, June 4–7, 2001, U.S. Department of Transportation, Washington, DC, United States.

Hackenberg, U., Rabe, M., and Friedewald, K. "Influence of Compatibility on Car Design," VDI-Berichte 1471, Innovativer Kfz-Insassen- und Partnerschutz, Tagung Berlin, September 30–October 1, 1999.

Schwarz, T. and Zobel, R. "Ermittlung der Zellsteifigkeit zur kompatiblen Auslegung von Pkw-Frontstrukturen," Crash-Tech 2000, May 18–19, 2000, München, TÜV Akademie GmbH.

Seiffert, U., Hamilton, J., and Boersch, F. "Compatibility of Traffic Participants," 3rd International Congress on Automotive Safety, Vol. 1, July 15–17, 1974, San Francisco, CA, U.S. Department of Transportation, United States.

Seiffert, U. "Höhere Aufprallgeschwindigkeit versus Kompatibilität," Tagung Crash Tech Special, TÜV Akademie GmbH, München, March 9–10, 1998.

Zobel, R., et al. "Development of Criteria and Standards for Vehicle Compatibility," 17th International Technical Conference on the Enhanced Safety of Vehicles, Amsterdam, June 4–7, 2001, U.S. Department of Transportation.

13-2. Klanner, W. "Status Report and Future Development of the Euro NECAP Program," Experimental Safety Vehicles (ESV) Conference 2001, Amsterdam.

13-3. Appel, H., et al. "Crash Compatibility of Passenger Cars Achievable But How?" Institution of Mechanical Engineers, London, UK.

13-4. BAST—Bundesanstalt für Straßenwesen, Germany, scientific report, undated.

13-5. Schwarz, T. "Selbst- und Partnerschutz bei frontalen Pkw-Pkw-Kollisionen (Kompatibilität)," Fortschritt-Berichte VDI Series 12, No. 502, VDI-Verlag, Düsseldorf, Germany, 2002.

13-6. Huibers, J. and de Beer, E. "Current Front Stiffness of European Vehicles with Regard to Compatibility," 17th International Technical Conference on the Enhanced Safety of Vehicles, Amsterdam, June 4–7, 2001, U.S. Department of Transportation, National Highway Traffic Safety Administration.

13-7. Relou, J. Methoden zur Entwicklung Crash-kompatibler Fahrzeuge, Dissertation, Technical University of Braunschweig, Shaker-Verlag, Aachen, Germany, 2001, ISBN 3-8265-7804-X.

13-8. Schimmelpfennig, K.-H. "Board Frame, a Possible Contribution to Improve Passive Safety," 15th International Technical Conference on the Enhanced Safety of Vehicles, Melbourne, Australia, May 1996, U.S. Department of Transportation, National Highway Traffic Safety Administration.

14.

Computer Support for the Development of Safety Components

14.1 The Basics

The virtual development process has already become a reality in the research and pre-development phases in all areas of the vehicle. This includes the area of accident avoidance and the field of mitigation of injuries. This chapter describes the tools for accident simulation. Together with the geometric data of the vehicle components, an optimization process among the design, performance during the accident, noise, and vibration must be performed. The breakthrough in the field of calculation for safety items was achieved by the development of stable software programs and the much higher performance of the supercomputer. One reason for the success was an interdisciplinary working group of the German research organization FAT [14-1]. Figure 14.1, created by Holzer et al. [14-2], shows how much progress in simulation techniques has been achieved in the meantime.

Independent of the success of the simulation technique, it is very important that the computation is supported by comparative analysis by using the results of hardware tests. The performance of the supercomputer has doubled nearly every two years, and many software programs running on computers can be used by engineers without limitations. Although the number of parameters that influence accident simulation tests has increased in recent years, calculation becomes an important part of the development process. In the newly defined product creation process, the simulation and calculation technique is an integrated part of marketing, development, production, finance, quality assurance, and service.

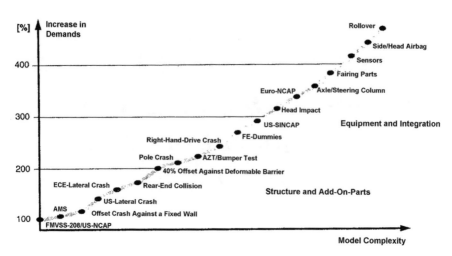

Figure 14.1 Simulation effort per car development versus model complexity. (Source: Ref. 14-2.)

14.2 Description of the Numeric Tools

The most important task for any computer-supported engineering work is the generation and use of physical–mathematical models.

During the creation of the model, you must know the physical input data and the vehicle component that shall be investigated. Depending on the status of the project, the results achieved by calculation will be able to support trends and to detail solutions up to full prognosis capability. A large number of devices are already available to convert physical models into mathematical ones.

For an automotive engineer, the development of new software is not as important as its creative use. For a crash calculation, several calculation models are used. These models include knowledge systems and a coupling of FEM and MKS models together with the test database. This database must be updated continuously. Today, material data, the connecting and the manufacturing process of the material also are part of the calculation method.

The CAD (computer aided design) database is the basis for the use of FEM (finite element model) calculation. Meanwhile, the performance of the total vehicle and the performance of single components are optimized by computer simulation techniques. Most automotive companies use both high-power supercomputers and simplified models that can run on less-expensive computers. In many applications, the performance of the vehicle structure is calculated by using FEM crash software with multi-body systems, such as MADYMO for the simulation of the occupant and restraint systems. The solution used today is a software-based coupling of both systems. This solution allows the replacement of certain modules and an integrated calculation of the total vehicle together with the restraint system (airbag and belts) and with the choice of different sizes of occupants.

It is important that the scientific knowledge database is updated continuously on the actual level for the optimization of certain vehicle components, as well as for total vehicle performance.

14.3 Calculation of Components

In many accidents, the front longitudinal beams are important vehicle parts for energy absorption. Numerous theories have been established by research in the area of fold bulging as one means to achieve an optimal energy conversion. In practical tests, the following formula for the average bulging force for the profile shown was determined, as shown in Table 14.1.

$$F = \overline{\theta}_F \cdot \delta_F \cdot \alpha_F \frac{S_x^2 \cdot U_x}{U_a}$$

where

$\overline{\theta}_F$ = Dimensional coefficient

δ_F = Elastic limit of the material

α_F = 1.0 to 1.5 velocity dependent

S_x = Sheet thickness

U_x = Profile circumference with the sheet metal thickness S_x

U_a = Total length of the profile

TABLE 14.1
CROSS SECTIONS OF DIFFERENT LONGITUDINAL BEAMS
(SOURCE: REF. 1-1.)

a mm	50 - 100	20 ÷ 100	50	70 a_1=80	110
h mm	-	20 ÷ 100	50 ÷ 70	110	20
f mm	-	10 ÷ 20	20	20	1,5
s_L mm	2,0 4,0	1,0 1,5	1,5 1,5	1,5	1,0
s_D mm	-	1,0 1,5	1,0 1,5	1,0	
θ_F	43 43	50 43	45 45	48	
s_θ	1,9 1,5	- 3,0	1,5 2,2	-	
	17 tests	$\dfrac{61\ \text{tests}}{\overline{\theta_F}=47,5}$ $s_\theta=4,7$	$\dfrac{9\ \text{tests}}{\overline{\theta_F}=45}$ $s_\theta=1,7$		$\dfrac{4\ \text{tests}}{\overline{\theta_F}=77}$

Figure 14.2 shows the results of FEM calculations of a longitudinal beam during the deformation phases.

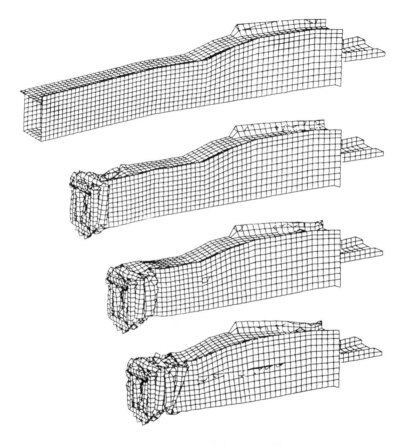

Figure 14.2 Finite element model (FEM) of longitudinal beams.
(Source: Ref. 1-1.)

The program module DYNA 3D and PAM-Crash can be used effectively in these types of calculation problems. For example, rectangular tubes of different profiles (square, hexagonal, and octagonal) were investigated by computer-supported calculation methods. The results demonstrate that hexagonal and octagonal tubes have a greater ability to absorb energy compared to that of square tubes. In real-world crash situations, the design of longitudinal beams and the transverse support in front of them are important in achieving the fold-bulging process of the beams. The fold-bulging process has the greatest capability for energy absorption during deformation. In the event of buckling of the longitudinal beams, the capability to convert the crash into deformation energy is the lowest. Figure 14.3 provides an overview of the calculation of deformation force versus crash time in a frontal collision for the front end of a vehicle.

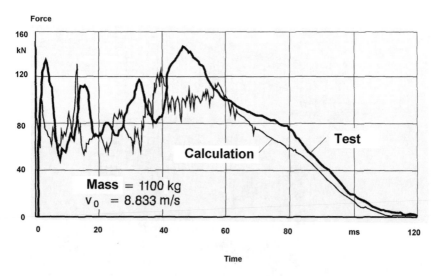

Figure 14.3 Comparison of deformation force as a function of time for the front end of a vehicle. (Source: Ref. 1-1.)

The calculation time for the crash simulation of a front end (not the total car) by the use of a modern supercomputer such as a NEC SX6 could be reduced to approximately 15 min.

The program PAM-STAMP [14.3] can even include the production process. Figure 14.4 shows how many parameters can be used today for this type of calculation. The program AUTOFORM simulates the material formation;

the effective plastic elongation is taken from the program PAM-STAMP. Together with the material data, we can calculate the foreseeable design.

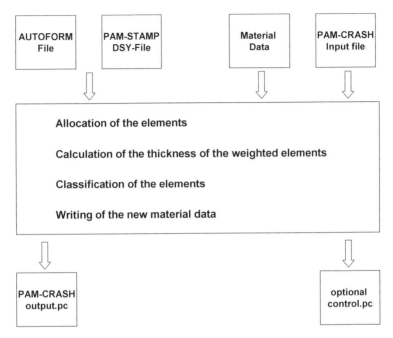

Figure 14.4 Data flow diagram for the simulation of mechanical data.
(Source: Ref. 14-3.)

14.4 Total Vehicle Crash Computation

In the concept study, the design engineer could run first calculations with specified key data, such as wheelbase, track width, powertrain location and size, auxiliaries, and other vehicle components. This allows the design engineer to perform a much better initial design study. Consequently, the first prototypes have much better quality with respect to their performance in crash tests. For the simulation tool, finite element model, of the total vehicle, 350,000 elements are used. This requires not only powerful computers (either parallel computers or a supercomputer) but good tools for generation of the finite element model. Figure 14.5 demonstrates the quality of the computer simulation technique in showing a 56 km/h (35 mph) offset crash against a deformable barrier.

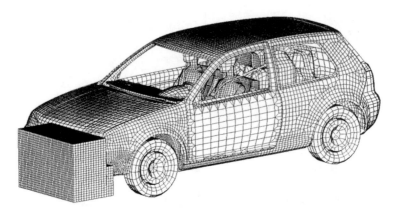

Crash beginning

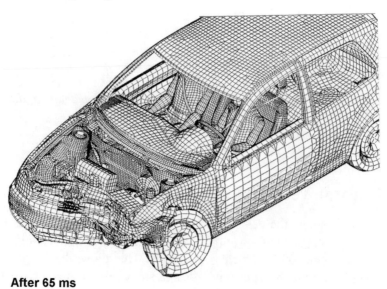

After 65 ms

Figure 14.5 Total car crash simulation.

The finite element model (FEM) could be used not only for the total vehicle crash but for the detailed analysis also. In a Ph.D. thesis, Hübler [14.4.] simulated the vehicle floor panel to determine whether vibrations could influence the triggering of the airbag sensors. In this case, the FEM for the body floor panel together with the longitudinal bars and the inner wheel-house had to use 668,000 elements to demonstrate the expected effect.

Through the level achieved today, the use of calculations became an important element in vehicle development. The necessary future activities for an even greater use is the consideration of the vehicle body structure together with the different types of occupants, restraint systems, and the interior of the vehicle. With this, the dummy, or with a more human-like dummy, performance in the vehicle environment could be better calculated.

14.5 Occupant and Restraint System Simulation

With the simulation of the dummies by the use of MKS (multi-body systems) and/or FEM, restraint system performance can be optimized. Using MKS as a dummy substitution, the necessary calculation time is much less, compared with the FEM dummy.

The MKS methods allow safety development primarily in three areas:

- Principle and trend analysis
- Plausibility of the system
- Components and detail development

A typical application field is the optimization of systems. Figure 14.6 shows one example. In this case, two different finite element models are used in a European side impact test. The results for the acceleration of the upper and lower spine are shown.

Meanwhile, MADYMO dummies of various sizes (through a child up to a 95% male) are used for accident simulation runs. This is possible even for out-of-position situations. Figure 14.7 shows the result of a calculation of an

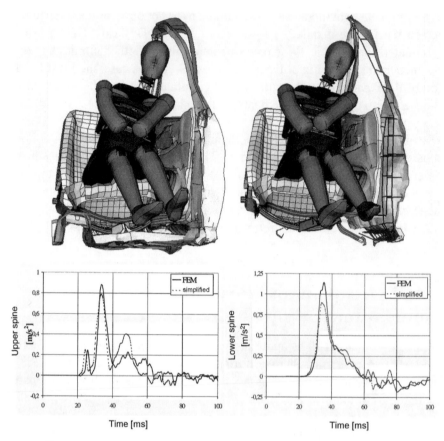

*Figure 14.6 Finite element models (FEM) for side impact and
acceleration of the upper and lower spine.*

unrestrained dummy test in the passenger side for the head and neck, compared to the measurement of two tests [14-5].

As we can see from the calculated and measured data, the match is sufficient.
In other cases, the trend of results could be reproduced but not the actual
numbers. Therefore, the development technology could be categorized into
the following groups:

•   Expanded MADYMO models with extensive use of finite element fea-
    tures, including a coupling with the PAM-Crash model.

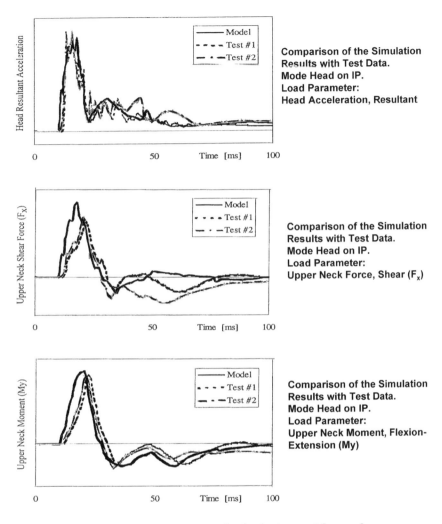

Figure 14.7 Comparison of calculations with test data.
(Source: Ref. 14-5.)

- Coupled MADYMO and PAM-Crash, where MADYMO is simulating the dummy and PAM-Crash is simulating the airbag and the vehicle.

- Using full finite element models (FEM) for occupant, airbag, and vehicle.

Special attention must be given to the simulation of side impacts. In a study of the optimization of vehicle stiffness distributions by means of numerical simulation for the development of crash compatible vehicle components by Kersten [14-6], it was found that the PAM-Crash Euro SID 1 FE-dummy gives a better representation of the dummy testing compared to the MADYMO Euro SID dummy.

14.6 Pedestrian Simulation Tests

Because of the importance of pedestrian protection and the on-going rule-making process, pedestrian simulation tests are done for the whole pedestrian as well as for component testing. Figure 14.8 shows a simulation of a pedestrian impact.

14.7 Summary

The simulation tools for evaluating the performance of the vehicle, its components, its occupants, and other traffic participants with respect to automobile safety have made a fundamental contribution to the product creation process. The continuous further development of these tools also in the areas of accident avoidance, software development, mechatronic systems, durability, and comfort is more than necessary to further improve the safety of traffic participants.

14.8 References

14-1. Raasch, I., Scharnhorst, T., and Schelkle, E. "Report of the FAT Working Group Finite Elements," ed. by Verband der Automobilindustrie (Association of the [German] Automobile Industry) (VDA), Frankfurt a.M., Germany, undated.

14-2. Holzner, M., et al. "The Virtual Crash Lab: Objectives, Requirements, and Recent Developments," in VDI-Berichte 1411, Numerical Analysis and Simulation in Vehicle Engineering, ed. by Verein Deutscher Ingenieure, VDI-Verlag, Düsseldorf, Germany, 1998, pp. 27–52.

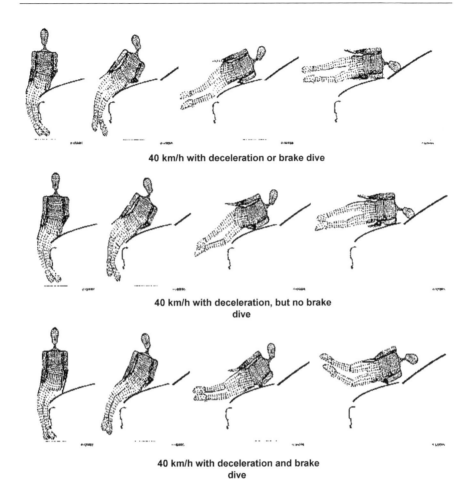

40 km/h with deceleration or brake dive

40 km/h with deceleration, but no brake dive

40 km/h with deceleration and brake dive

Figure 14.8 Simulation of a pedestrian impact against the front end of a passenger car traveling at 40 km/h (25 mph). (Source: Ref. 14-7.)

14-3. Scholz, S.-P., et al. "Crash Computation in Consideration of the Metal Forming Process," in VDI-Berichte 1411, Numerical Analysis and Simulation in Vehicle Engineering, ed. by Verein Deutscher Ingenieure, VDI-Verlag, Düsseldorf, Germany, 1998, pp. 195–214.

14-4. Hübler, R. Unterstützung bei der Auslegung von Airbagsystemen durch FEM-Berechnung, M-Verlag, Mainz, Germany, 2001, ISBN 3-89653-828-4.

14-5. Siebertz, K., et al. "Occupant Protection Assessment from Out-of-Protection Models," in VDI-Berichte 1411, Numerical Analysis and Simulation in Vehicle Engineering, ed. by Verein Deutscher Ingenieure, VDI-Verlag, Düsseldorf, Germany, 1998, pp. 111–132.

14-6. Kersten, R. "Optimization of Vehicle Stiffness Distributions," internal scientific report, University of Eindhoven, The Netherlands, 2002.

14-7. Koch, W., et al. "Comprehensive Approach to Increased Pedestrian Safety in Pedestrian-to-Car Accidents," in VDI-Berichte 1637, Innovative Occupant and Partner Crash Protection, ed. by Verein Deutscher Ingenieure, VDI-Verlag, Düsseldorf, Germany, 2001.

14-8. Verein Deutscher Ingenieure (Ed.), Numerical Analysis and Simulation in Vehicle Engineering (1998), VDI-Berichte 1441, VDI-Verlag, Düsseldorf, Germany, 1998, See also Verein Deutscher Ingenieure (Ed.), Numerical Analysis and Simulation in Vehicle Engineering (2000), VDI-Berichte 1559, VDI-Verlag, Düsseldorf, Germany, 2000.

Index

About the Authors

Professor Dr.-Ing. Ulrich Seiffert currently is the acting chairman of WiTech Engineering GmbH, speaker of the Center of Traffic Management, and a lecturer at the Technical University of Braunschweig since 1982. Until the end of 1995, he was a member of the board for research and development of Volkswagen AG in Germany. He also holds a number of patents in the safety area and was awarded several times for his work in the field of safety engineering. He is author of a large number of books and has published more than 100 technical papers.

Dr.-Ing. Lothar Wech is the general manager of TÜV Automotive GmbH, TÜV Süddeutschland Group. He is competent in the field of passive safety of vehicles and has served as chairman of the Crash-Tech Conferences since 1992.